GET READY FOR

A&P for Nursing and Healthcare

Visit the *Get Ready for A&P for Nursing and Healthcare* Companion Website at **www.pearsoned.co.uk/getready** to find valuable **student** learning material including:

- Interactive tutorials for you to study, and then test yourself
- Quizzes and animations to bring the subjects to life
- Flashcards to help you with your revision
- Links to relevant sites on the web

GET READY FOR

A&P for Nursing and Healthcare

Lori K. Garrett
Danville Area Community College

UK ADAPTATION

Ailsa Clarke BSc, MSc, PGCEA, FHEA
former Lecturer in Biosciences,
European Institute of Health and Medical Sciences,
University of Surrey

and

Pearl Shihab BSc, MSc, RN, RM, RNT
former Senior Tutor in Nursing Studies,
European Institute of Health and Medical Sciences,
University of Surrey

Harlow, England • London • New York • Boston • San Francisco • Toronto
Sydney • Tokyo • Singapore • Hong Kong • Seoul • Taipei • New Delhi
Cape Town • Madrid • Mexico City • Amsterdam • Munich • Paris • Milan

Original US edition: *Get Ready for A&P*
Editor-in-Chief: Serina Beauparlant
Development Manager: Claire Alexander
Associate Media Editor: Ryan Shaw
Assistant Editor: Jessica Brunner
Market Development Manager: Erin Joyce
Managing Editor/Production Supervisor: Wendy Earl
Cartoonist/Cover Artist: Kevin Opstedal
Text Designer: Seventeenth Street Studios
Cover Designer: Yvo Riezebos
Photo Researcher: Travis Amos
Art Coordinator: David Novak
Compositor: Seventeenth Street Studios
Senior Manufacturing Buyer: Stacey Weinberger
Executive Marketing Manager: Lauren Harp

UK adaptation: *Get Ready for A&P for Nursing and Healthcare*
Acquisitions Editor: Kate Brewin
Desk Editor: Philippa Fiszzon
Text Designer: Colin Reed
Cover Designer: Andrea Bannuscher
Copy Editor: Rose James
Production Controller: Kay Holman
Proofreader: Robert Chaundy
Permissions: Mavis Collins
Marketing Manager: Santiago Ruiz De Velasco

Learning Styles Assessment on pages 5–7
© Marcia L. Conner, www.agelesslearner.com

Pearson Education Limited
Edinburgh Gate
Harlow
Essex CM20 2JE
England

and Associated Companies throughout the world

Visit us on the World Wide Web at:
www.pearsoned.co.uk

Authorised adaptation from the United States edition, entitled GET READY FOR A&P, 1st Edition, ISBN: 0805382844 by GARRETT, LORI K., published by Pearson Education, Inc., publishing as Benjamin Cummings, Copyright © 2007

First adaptation edition published by Pearson Education Limited, Copyright © 2008

ISBN: 978-0-273-71360-9

British Library Cataloguing-in-Publication Data
A catalogue record for this book is available from the British Library

10 9 8 7 6 5 4 3 2 1
11 10 09 08 07

Typeset in 11/15 pt Minion by 73
Printed in Great Britain by Henry Ling Ltd, at the Dorset Press, Dorchester, Dorset

The publisher's policy is to use paper manufactured from sustainable forests.

Contents

Visit the *Get Ready for A&P for Nursing and Healthcare* Companion Website at **www.pearsoned.co.uk/getready** to find valuable **student** learning material including:

- Interactive tutorials for you to study, and then test yourself
- Quizzes and animations to bring the subjects to life
- Flashcards to help you with your revision
- Links to relevant sites on the web

Preface to *Get Ready for A&P*

Welcome to the fascinating world of anatomy and physiology! Most students take A&P because they have to – it's a required part of their educational curriculum. I hope you will quickly discover how amazing the human body is and become intent on learning all that you possibly can about it. After all, you *are* one!

For many reasons, students sometimes do not succeed in their first anatomy and physiology class. If you're reading this Preface, you probably have a strong desire to succeed, and you know the competition for admission into numerous educational programmes is increasing. You are probably keenly aware that you can't just pass your classes – you need to do quality work and truly master the course content. This book offers you the opportunity to enhance your performance in this rigorous course. It is designed to help you prepare to be successful.

You may be using this book before your course officially begins. Your lecturer may have assigned it to you as homework during the first week or two of classes. Perhaps you are using it on your own and will come back to it periodically throughout the year as you move along your programme. However you use it, the purpose of the book remains the same. The goal is to help you get a strong start in anatomy and physiology, and to master the material not just for exams, but for your future as well. *Get Ready for A&P* contains six relatively short and interactive chapters that engage you every step of the way. You'll read, but you'll also frequently do activities.

The book starts with basic study skills in **Chapter 1**. As with any course, you will get out of A&P what you put into it, and it will take significant time and effort on your part to succeed. This chapter helps you focus and manage your time so you can first find time to study, then use your study time effectively. After exploring different learning styles you can assess which style best fits you, then discover specific study strategies that complement your preferred style. You'll assess your current habits as a student and learn specific tips and strategies to help you study better. Specific tips will help you write your notes, read textbooks and take tests.

Chapter 2 covers basic maths skills. Anatomy and physiology are sciences, and all science involves at least some maths. This chapter takes you from basic maths operations through reading and interpreting numerical information in graphs and tables – the maths you'll need for a head start in your course.

Many of the words in your A&P class will sound foreign to you, and well they should! Most of the terms come from Latin or Greek. Knowing the terms underlies all aspects of learning in this class. In **Chapter 3**, Terminology, we look at how the words are built and learn some simple tricks that will rapidly expand your A&P vocabulary and have you talking like a pro!

The second half of the book gets more specific and parallels some of what you will cover in the first few chapters of your A&P textbook. In **Chapter 4**, we cover body basics: some general biological principles that guide how the body works, and a quick overview of each organ system. You'll begin to understand how your body is arranged and how the different systems function.

In **Chapter 5**, we tackle some basic chemistry. This chapter gives you the basics, from atoms to organic molecules. We discuss some tricks for gaining information from the periodic table, and see how atoms join together to form molecules. If you can understand bumper cars, you can understand bonding!

Finally, in **Chapter 6**, we explore cells. We are made of trillions of them, and almost everything that happens in our bodies occurs inside our cells. We discuss basic cell structure, the cell life cycle and cell reproduction.

Now that you know the road map for this book, let's explore the stops you'll find along the way. Here are the special features in each chapter, designed to keep you involved and to make you a better student in A&P:

■ *Your Starting Point* tests your grasp of the chapter content before you start. Answers are provided for all of these except in Chapter 1, where the answers are personal.

■ *Quick Check* asks you to recall or apply what you have just read, to keep your eyes from scanning the page while your brain is on holiday. The answer is provided on the same page.

■ *Picture This* asks you to visualise scenarios and then answer questions about them, to help you better understand the topics.

■ *Time to Try* is a simple experiment or quick assessment in which you perform an active exercise.

■ *Why Should I Care?* highlights the relevance of the material so you understand its importance in the big picture.

■ *Reality Check* assesses whether you really understood the material.

■ *Applying the Theory* uses examples of actual situations which link directly with the chapter content. It's like mini case studies.

■ *Look Out* paragraphs highlight possible pitfalls or challenges to a student. They focus on areas where a novice practitioner may not understand the consequences of a particular action or inaction.

■ *Keys* highlight main themes or statements for reinforcement and easy review.

■ *Running Words* list key terms from the chapter to help you start your own running vocabulary list by writing each term in a notebook, then defining it.

■ *What Did You Learn?* end-of-chapter quizzes may include short answer, multiple choice, or matching exercises. The answers appear at the end of the book.

■ *Web Resources* provide links to useful Internet sites at which you can learn more or practice what you've learned. Many of these provide additional activities or quizzes as well.

Finally, check out the online component of *Get Ready for A&P* at **www.pearsoned.co.uk/getready**, where you can study and quiz yourself with interactive tutorials, quizzes, animations, flash cards and an audio glossary.

I wrote *Get Ready for A&P* in a conversational tone because that is how I teach and because science is too often made boring by boring presentation. Science shouldn't be stuffy – it should be fun! Learning should be peppered with giggles, salted with silliness and dotted with *AhHA!* moments. Sometimes it seems the terminology alone can put you in a trance, so why should the style?

Now it is time to dig in. So get comfortable, and *Get Ready for A&P*!

Lori K. Garrett

Adaptors' Preface to *Get Ready for A&P for Nursing and Healthcare*

We were excited to discover *Get Ready for A&P*, which had a format and content very suited to our students, who come from a wide variety of educational backgrounds. We recommended it to students as a text to buy and study before starting their chosen healthcare programme: once they had started their course it was used to bridge the gap between their knowledge and the more advanced textbook.

Some modifications were necessary to make the content more accessible to UK students, easier to understand and more applicable to their work in the UK by using metric units and more familiar terminology. We hope the book will help students appreciate the reasons for studying the theory behind the biomedical sciences that may seem unrelated to their vision of what they need to know in order to care for patients.

With this in mind, the text has been adapted using English grammar and spelling, particularly relevant to developing your vocabulary. Metric units have been used throughout, with examples of calculations used in practice. The special features in each chapter which were designed to keep you involved and make you a better student in A&P have also been linked to the UK experience. In addition, we have added a feature called *Applying the Theory* which uses examples of actual situations which directly link with the chapter content. Areas where a novice practitioner may not understand the consequences of a particular action or inaction are also put into context in the *Look Out* paragraphs.

The **Web Resources** have been updated and related to UK websites where appropriate. In this way, after reading the text in the book you can look at some up-to-date interactive resources to reinforce your learning and help you learn in a different way. Many students prefer moving diagrams to just reading the words on a page.

In addition, the text in the book has now been linked to the website designed especially to support your learning. Click on the link: **www.pearsoned.co.uk/getready**. With this online component of

Get Ready for A&P for Nursing and Healthcare you can study and quiz yourself with interactive tutorials, multiple choice questions, animations, flash cards and an audio glossary. The book has links to these activities at the appropriate moment in the text to reinforce your learning and help your understanding.

This has been a very exciting project for us all. We hope you find it beneficial for your studies and indeed throughout your life-long learning once you have qualified.

Ailsa Clarke
Pearl Shihab

Acknowledgements

Ah, where to begin? This project began with what seemed a harmless email from Assistant Editor Jessica Brunner, asking if I would consider submitting a sample chapter for a new workbook. I took the bait. It has been a whirlwind of frenzied activity from that moment, and never have I worked on a project for which the team was more important. To make this book possible, an amazing group pulled together, taking on lots of 'other duties as assigned', and ignoring conventional wisdom. I thank Wendy Earl for coming up with a new 'Production Paradigm' to make this book happen on an incredibly accelerated schedule, for her rapid turnarounds on copy editing, and for developing a schedule we, surprisingly, found we could almost live with! I thank Randall Goodall for designing the text and keeping it user-friendly, and for so kindly making all our picky revisions. I thank all the gang with him at Seventeenth Street Studios, especially Richard and Valerie, for their work on the paging and art. I thank Kevin 'Kartoon' Opstedal for his wonderful cartoon illustrations that kept this book fun, the way science *should* be. And I thank Travis Amos for so quickly getting us just the right photos.

I especially thank the editorial team at Pearson Benjamin Cummings. Jessica, thanks for making that initial contact, for all the shipments you sent to me, for being the one left behind when others were travelling, and for all the assorted details you tended to both in and out of my direct line of fire. I wholeheartedly thank Claire Alexander. Claire – I chuckle as I think of you now! I will miss our daily laughs, the lengthy phone chats, and the silliness that kept us both going. And I will always be grateful for the coffee beans – a guaranteed way to help meet tight deadlines. And with greatest respect, I thank Serina Beauparlant for remembering me from previous work and giving me this super opportunity. Serina, you have always made me feel invaluable and been remarkably accommodating. I truly appreciate how closely you followed this project, staying in touch from the road, checking in regularly just to be sure I was happy, and giving me such freedom. I especially thank you for recognising the need for this project, valuing my skills and knowledge, and pulling together this fantastic and fun team that worked against all odds and that ever-ticking clock to do the impossible. Thanks for my best publishing experience ever.

I also want to extend special thanks to Judy Megaw of Indian River Community College, who took charge of the online component, ensuring a robust website that fully supports this book.

I would be remiss if I didn't thank the many reviewers, listed on the following page, who carefully considered these chapters and gave me their candid feedback – the book is stronger because of their input.

Finally, I gladly acknowledge the home team here at Danville Area Community College. I thank my students for being so supportive and for doing their jobs well so I look good in mine. I thank Donna Davis for getting all those last-minute overnight shipments out the door. I thank Dean Janet Redenbaugh for her constant support of me and all my undertakings, and I thank John Hoagland for putting up with all the insanity and chaos.

Lori K. Garrett

REVIEWERS

Erin Amerman
Santa Fe Community College

Vince Austin
Bluegrass Community & Technical College

Claudie Biggers
Amarillo College

Margaret Creech
Laramie County Community College

Terry Harrison
Arapahoe Community College

Clare Hays
Metropolitan State University

Maurice Heller
El Paso Community College

Mark Hubley
Prince George's Community College

Catherine Hurlbut
Florida Community College, Jacksonville

Jody Johnson
Arapahoe Community College

Johanna Kruckeberg
Kirkwood Community College

Ken Malachowsky
Florence-Darlington Technical College

Elaine Marieb
Holyoke Community College

Judy Megaw
Indian River Community College

Claire Miller
Community College of Denver

Amy Nunnally
Front Range Community College

Wayne Seifert
Brookhaven College

Alan Spindler
Brevard Community College

Dieterich Steinmetz
Portland Community College

Yong Tang
Front Range Community College

Deborah Temperly
Delta College

Jennifer van de Kamp
Front Range Community College

UK ADAPTATION

The publisher of this UK adaptation thanks the following reviewers for their valuable comments and feedback:

Martin Steggall
City University, London

Gay James
Coventry University

Peter Bentley
City University, London

Oonagh McNally
University of Ulster

Pat Pass
University of Greenwich

Sue Harry
University of East Anglia

Peta Reid
University of Southampton

Lazar Karagic
De Montfort University

1 Study Skills

The Proper Care and Feeding of a Human Brain

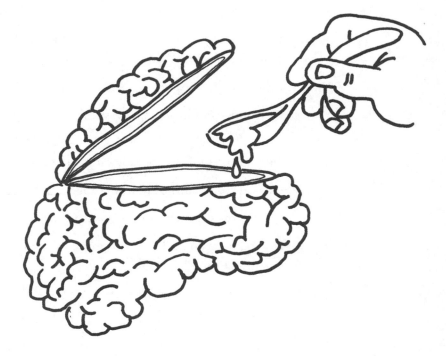

When you have completed this chapter, you should be able to:

- Understand your preferred learning style and the study strategies that support it.

- Have a set of skills that will help you get the most benefit from lectures and your reading.

- Have a written schedule that includes adequate study time.

- Know how to prepare well for an exam.

- Understand that you are ultimately accountable for your own success or failure.

- Feel ready and more confident to start your course.

YOUR STARTING POINT

Answer the following questions to assess your study habits.

1. How often do you read a course textbook? ____4____

2. How many days of the week do you study for one course? __4__

3. Do you study hard the day before an exam, but rarely between exams? __No__

4. Where do you study? __Home / uni__

5. How long should you spend studying outside of the classroom? __As much as poss__

6. Do you schedule your study time and stick to it? __No__

7. Do you study hard or hardly study? __Hard__

8. Do you mostly memorise when studying for a test? __Try to__

9. Do you have a good support group of family and friends who encourage you? __Yes__

10. Do you quiz yourself when studying? __No__

Welcome to the exciting and sometimes challenging world of anatomy and physiology! You will quickly discover how amazing the human machine truly is – a curious marvel of complexity that is simultaneously surprisingly simple. You will be fascinated by learning how your own body is built (anatomy) and how it works (physiology), and interest in your subject matter always makes it much easier to learn.

Still, no matter how exciting your anatomical explorations may be, your course may, at times, seem rigorous and demanding. You've taken a great first step by turning to this book to jump-start your studies. This book is meant to help you enter the course with a well-planned strategy for success and with confidence in your basic science knowledge.

Answers: Answers will be individualised, except for number 5 – you should spend 2–3 hours studying for each hour of class time.

TIME TO TRY

Are you wondering why you are reading this book?

You have committed yourself to studying science as part of your course. Do you think this is going to be difficult?

Go to *Get Ready for A&P for Nursing and Healthcare* on your computer. Enter the web address **http://www.pearsoned.co.uk/getreadyquiz**

Try to complete the quiz. This will help you to see what kind of computer skills you need to be able to study online and also test you on your preliminary knowledge.

The purpose of this chapter is to help you 'train your brain' to make your learning process easier and more efficient.

Why Should I Study Anatomy and Physiology?

Most students take anatomy and physiology because it is required for their chosen career pathway. Sometimes when something is required, we do it only because we have to, without considering what benefits the task might hold for us. Unfortunately, some students use that approach for anatomy and physiology, but it is always easier to study something if you understand why it matters, and this course is no exception.

PICTURE THIS

Until recently your car has run perfectly, but now the engine occasionally stops running and is difficult to restart. Assuming you have little knowledge of car mechanics, you are not likely to solve the mystery or make repairs yourself. You take it to a car mechanic, who will consider how your car is malfunctioning – its symptoms, if you like – and then fix it. What knowledge will the mechanic need to accomplish that goal? _____

_____

_____

In what ways are people in health- and medical-related fields similar to the car mechanic? _____

_____

Why do they need to fully understand anatomy and physiology?

_____

_____

Now consider your own future – what is your planned career?

_____

Why will you need to know anatomy and physiology?

_____

Many anatomy and physiology students plan careers in a medical or health field. Others may be heading into physiotherapy, dietetics, sports science, perhaps biomechanics or bioengineering, and many other fields. These career areas share a common thread – anatomy and physiology form the foundation on which they are all built. Now, back to our example. To understand your malfunctioning car, the mechanic must first fully understand the parts of your car – how they fit together and how they normally function, just as you will need to understand the parts of the human body and their normal functions. Finally, there is a simpler reason why you should care about learning anatomy and physiology. The human body is an amazing machine, and you own one. Anatomy and physiology are your owner's manual.

To Thine Own Self Be True: **Learning Styles**

What *is* the best way to learn these subjects? A huge amount of research has explored how people learn, and there are many opinions. One common approach considers which of the senses a learner relies on the most – sight, sound or touch:

- Visual learners learn best by *seeing*, for example, watching a demonstration.

- Auditory learners learn best by *hearing*, for example, hearing an explanation.

- Tactile (kinaesthetic) learners learn best by *doing*, for example, putting your hands on something and feeling it.

TIME TO TRY

Let's uncover your learning style.

1. Look at **Table 1.1**. Read an activity in the first column, then read each of the three responses to the right of that activity.

2. Mark the response that seems most characteristic of you.

3. After doing this for each row, you are ready to total up your score. Add all the marks in each column and write the total in the corresponding space in the bottom row.

4. Next look at your numbers. You will likely have a higher total in one column. That is your primary learning style. The second-highest number is your secondary learning style.

My primary learning style is: _____

My secondary learning style: _____

TABLE 1.1   Assessing your learning style.

Activity	Column 1	Column 2	Column 3
1. While I try to **concentrate** . . .	I grow distracted by clutter or movement, and I notice things in my visual field that other people don't.	I get distracted by sounds, and I prefer to control the amount and type of noise around me. ✓	I become distracted by commotion, and I tend to retreat inside myself.
2. While I am **visualising** . . .	I see vivid, detailed pictures in my thoughts. ✓	I think in voices and sounds.	I see images in my thoughts that involve movement.
3. When I **talk to someone** . . .	I dislike listening for very long.	I enjoy listening, or I may get impatient to talk.	I gesture and use expressive movements. ✓

TABLE 1.1   Assessing your learning style, continued.

Activity	Column 1	Column 2	Column 3
4. When I **contact people** . . .	I prefer face-to-face meetings.	I prefer speaking by telephone for intense conversations.	I prefer to interact while walking or participating in some activity. ✓
5. When I **see an acquaintance** . . .	I tend to forget names but usually remember faces, and I can usually remember where we met.	I tend to remember people's names and can usually remember what we discussed.	I tend to remember what we did together and may almost 'feel' our time together. ✓
6. When I am **relaxing** . . .	I prefer to watch TV, see a play, or go to the cinema. ✓	I prefer to listen to the radio, play music, read, or talk with a friend.	I prefer to play sports, do crafts, or make something with my hands.
7. While I am **reading** . . .	I like descriptive scenes and may pause to imagine the action.	I enjoy the dialogue most and can 'hear' the characters talking. ✓	I prefer action stories, but I rarely read for pleasure.
8. When I am **spelling** . . .	I try to see the word in my mind or imagine what it would look like on paper.	I sound out the word, sometimes aloud, and tend to recall rules about letter order.	I get a feel for the word by writing it out or pretending to type it. ✓
9. When I **do something new** . . .	I seek out demonstrations, pictures or diagrams.	I like verbal and written instructions, and talking it over with someone else. ✓	I prefer to jump right in to try it, and I will keep trying and try different ways.

TABLE 1.1 Assessing your learning style, continued.

Activity	Column 1	Column 2	Column 3
10. When I **assemble something** . . .	I look at the picture first and then, maybe, read the directions.	I like to read the directions, or I talk aloud as I work. ✓	I usually ignore the directions and figure it out as I go along.
11. When I am **interpreting someone's mood** . . .	I mostly look at their facial expressions.	I listen to the tone of the voice.	I watch body language. ✓
12. When I **teach others how to do something** . . .	I prefer to show them how to do it.	I prefer to tell them or write out how to do it.	I demonstrate how it is done and ask them to try. ✓
TOTAL:	**Visual:** 2 / 4 / 2	**Auditory:** 4 / 3 / 3	**Tactile/Kinaesthetic:** 6 / 5 / 7

Source: Courtesy of Marcia L. Conner, www.agelesslearner.com

Now that you know your primary and secondary learning styles, you can design your study approach accordingly, emphasising the activities that use your preferred senses. Look closely at your scores, though. If two scores are rather close, you already use two learning styles well and will benefit from using both of them when studying. If your high score is much higher than your other scores, you have a strong preference and should particularly emphasise that style. Most people use a combination of learning styles.

In addition, information coming in through different senses reaches different parts of your brain. The more of your brain that is engaged in the learning process, the more effective your learning will be, so try strategies for all three styles and then emphasise your preferred style over the others. You'll know which strategies work best for you. Some strategies that you might try are summarised for you in **Table 1.2**.

TABLE 1.2   The three learning styles and helpful techniques to use in your studies.

 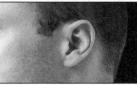 

	Visual	Auditory	Tactile
Techniques to use	❏ Sit close to the teacher. ❏ Take detailed notes. ❏ Draw pictures. ❏ Make flow charts. ❏ Use flash cards. ❏ Focus on the figures, tables, and their captions. ❏ Try using colouring books and picture atlases. ❏ Use visualisation.	❏ Listen carefully to your teacher's voice. ❏ Read the textbook and your notes out loud. ❏ Record lectures and listen to them later. ❏ Listen during the class instead of writing notes. ❏ Work in a study group. ❏ Discuss the material with others.	❏ Highlight important information while reading. ❏ Write your own notes in class and while reading the textbook. ❏ Transfer your notes to another notebook or type them into your computer. ❏ Doodle and draw as you read. ❏ Build models of anatomical structures. ❏ Create and conduct your own experiments. ❏ Hold your book while reading. ❏ Walk or stand while reading. ❏ Use anatomy colouring workbooks. ❏ Use flash cards.

Visual Learners

If you are a **visual learner**, you rely heavily on visual cues. You notice your teacher's mannerisms, expressions, gestures and body language. Seeing these cues is especially helpful, so sit at the front of the classroom, close to the teacher. You tend to think in pictures and learn well from visual aids such as diagrams, illustrations, tables, animations, film clips and handouts. Here are some strategies for you.

- In class, take detailed notes and make sketches.

- When studying on your own, draw pictures that relate to the information, make flow charts and concept maps, use flash cards, focus on the illustrations and tables in your textbook, and read the captions that accompany them.

- Use anatomy and physiology colouring workbooks and online tutorials with animations.

- Use mental visualisation of the material you are studying and imagine yourself acting out processes. For example, to learn major blood vessels, you might imagine yourself swimming through them.

Auditory Learners

If you are an **auditory learner**, you learn well from traditional lectures and discussion. You listen carefully to your teacher's vocal pitch, tone, speed and mannerisms. Material that you struggle with while reading becomes clearer when you hear it. Here are some strategies for you.

- Read the textbook and your notes out loud.

- Record the lectures so you can listen to them later. Recording lectures also allows you to listen during class instead of focusing on writing, which is less beneficial for you.

- Work in a study group, and discuss material with your teacher friends and colleagues.

Tactile Learners

If you are a **tactile learner**, you learn best by actively participating and doing hands-on activities. You may become bored easily in class from sitting still too long and start fidgeting or doodling. You need to do something physical while studying and learning. Here are some strategies for you.

- Try using a marker to highlight important information while you are reading.

- Write out your own notes in class and while reading the textbook. Later, transfer your notes to another notebook or type them into your computer.

- Draw pictures of appropriate material as you read.

- Build models of anatomical structures using clay or other materials.

- Create and conduct your own experiments.

- Hold your book and walk while reading.

- Use anatomy and physiology colouring workbooks.

- Make and use your own flash cards.

- Keep your hands and your mind busy at the same time.

LOOK OUT

Understanding your own learning style allows you to develop more effective and efficient study techniques that take advantage of your sensory preferences. Emphasising your preferred learning style will make the material easier to learn and it will stay with you for longer.

✔ <u>QUICK CHECK</u>

Homemade flash cards would be most beneficial to which two learning styles? _____ and _____.

How could they be used to benefit a learner of the third style?

_____

Answer: They would benefit visual and tactile learners. Reading them out loud would benefit auditory learners.

Putting on Your Best Face: **Getting Ready**

Many students mistakenly wait until the first lecture to start thinking about what it's like going to college. The key to starting your course well is to be organised and ready when you enter the classroom. This takes advance planning, but the time invested will save you even more time when the course is underway.

Putting it in Writing

As the course begins – if not before – you should get organised, and that begins with making a commitment to yourself. Too often we begin a project without setting goals in advance. If you set a goal, you enter with a purpose and a direction. If you do not set a goal, it's too easy to just go along and see what happens. Take time to think about your goals for the term. They should be both specific and attainable. Be realistic. For example, it may not be realistic to set the goal of always having the highest score in class, but a goal of getting an A in the class might be attainable. Once you've decided on your goals, write them down to give them more importance. Once you've written them, be firmly committed to them. To reinforce these goals, write them on an index card and place it in a prominent location in your study area so you'll see them every day.

TIME TO TRY

Set three main goals for yourself in this class, and write them below. Explain why achieving each goal is important to you.

Goal 1: _____

It is important to me because: _____

_____

Goal 2: _____

It is important to me because: _____

_____

Goal 3: _____

It is important to me because: _____

_____

Putting it all Together

I'm amazed when students show up for a test with no writing equipment! Don't let that happen to you. The more organised you are, the more efficient you will be, so let's organise what you will need for class. Categorise the items by what you take to class every day, what remains at home in your study area, and optional items that are nice, but non-essential, additions. Use the checklist provided for you in **Table 1.3**. Search your house and you'll probably find that you have many of these items already. Most of them can be bought at your college bookshop, but many are available at your local shops. We will discuss some of these items specifically.

You'll be going back and forth to class a lot, so it is most efficient to keep all the items you might need for class in one place. To carry them, most students use a backpack or briefcase. One advantage to this is that you can load it up with the essentials so that they are always ready when you leave for college. Let's discuss some of the items to pack.

You need a pocket-sized diary that has plenty of room for writing and that you can keep with you at all times. Or you may opt for a personal organiser portfolio or an electronic organiser. Select one you like, because you'll use it every day. In it, write all important dates you already know – when classes begin, holidays, when exams begin. Enter

TABLE 1.3   Organiser's checklist.

Item	✎✗
To take to class each day:	
Backpack/briefcase/rolling carrier	
Textbooks/workbooks	
Pocket-sized diary	
To Do list	
Separate notebooks for each course	
Copy of class schedule with buildings and room numbers	
Several blue or black ink pens	
Several pencils	
Small pencil sharpener	
2–3 coloured highlighter pens	
Small stapler	
Supply of paper for note-taking	
Calculator	
At home:	
Master calendar	
Separate file or folder for each course	
Loose notebook paper	
Index cards for making flash cards	
Computer paper for printer	
More writing utensils (pens and pencils)	
Stapler	
Calculator	
Scissors	
Paper clips	
Optional:	
Personal organiser	
Colouring workbooks associated with textbook	
Coloured markers/pencils	
Recorder to record lectures/readings	
A means of digital storage	
Extra batteries	
Anatomy atlas	
Medical dictionary	

your college timetable, your work schedule, and any other known time commitments. Try to keep your diary current so you always know how your time is being spent and can plan ahead.

If it's not part of your diary, you need a separate To Do list. Write all assignments and due dates on this list. You want one single To Do list for all of your classes as well as non-school activities, because they must all be done from the same pool of time. Writing them down allows you to view the entire list and review the deadlines for each item so you can easily prioritise, doing the assignments in the order in which they are due.

Maintain a record of all grades you receive (**Figure 1.1**). For each graded item, list what it is, when you handed it in, when you got it back, how many points you received, how many points were possible, and any additional notes. Once you know how your grade will be determined for the course, you can use this to keep track of your progress as you go along. It also provides a back-up in case there is any confusion later about your grade or your work.

Check with your lecturers to see if you should bring your textbook to the lesson. Typically, you may not need it in a lecture, but you may need to refer to it during the day.

Graded item	Date turned in	Date returned	My score	Possible points	Notes
Lab 1	9/6	9/13	10	10	Worked with Emily, Mike, Tom
Quiz 1	9/7	9/9	18	20	Study terms again
Lab 2	9/13	9/16	6	10	Messed up the maths!
Pop Quiz	9/14	9/16	5	5	From yesterday's lecture.
					I was ready!
Quiz 2	9/21	9/25	19	20	Forgot to answer one question!

FIGURE 1.1   A sample grade record for keeping track of your progress.

Always carry the basics with you. You need a notebook for note-taking. If you come to a lesson without a writing utensil, it says you do not think that anything said is important enough to write down. If you ask for a stapler before handing in an assignment, it says you threw it together with little thought. Pens, pencils, erasers, staples, paper, highlighters, coloured pencils, index cards, paper clips – these are just some items you may find useful. Replenish your supply as needed.

Set up your home study space like a home office. Be sure to have all the essential office supplies on hand – plenty of writing utensils, paper, a stapler, and so on. A critical part of the home study area is the master calendar. There are large desktop versions and wall charts, for example. You could use a calendar feature on your computer, but the more visible the calendar is, the more often you will look at it. This calendar should be large enough to accommodate plenty of writing, so think BIG! Each day, you should add anything that you put in your diary or on your To Do list to this master calendar. All time commitments should be entered, so also add all personal appointments and holidays. This is how you will schedule your life while in college, and the practice will probably stick with you far beyond that.

You may have already started your course before using this workbook. You certainly cannot do all these things before class at this point, but it is never too late to get organised. So, go and get it all together!

 LOOK OUT

To be successful in class, your effort should start before class begins. What are some tasks you should do before the first day of class?

_____

_____

_____

Set and write down your goals, organise the items you will need for class and at home, pack your bag, start your diary and master calendar, and organise your study area.

I Hate to **Lecture** on this, but Can You Hear Me Now?

Welcome to class! Imagine that it is the first day. You walk into class.

Where do you sit? _____

Why do you sit there? _____

The best seat in the house is front and centre. Obviously not everyone can sit there, but you should arrive early enough to sit within the first few rows and as near to the middle as possible. You want an unobstructed view of the teacher and anything they might show, because anatomy and physiology are very visual subjects. People sitting on the sides or in the back often do not want to be called on, or they want to be in their own space. They are not very engaged in the class. Don't let that be you. To succeed, you need to focus all of your attention on your teacher, minimise distractions, and participate actively. Lecturers tend to teach to the middle of the room (**Figure 1.2**). In fact, if your teacher is right-handed and uses equipment, such as an overhead projector, that is positioned on the right, their focus shifts to their right. You want to see your teacher and you want your teacher to see that you are present, listening and actively engaged.

Some teachers provide lecture notes so you can sit back and really think about what is being said. Notes or not, you need to get all the information you can from each lecture. Remember your learning style and use techniques that enhance it. We will discuss note-taking later, but consider recording the lectures. That way you miss nothing, and you can listen to the recording repeatedly, replaying it as needed. Another good technique is to write out your own notes while listening to the recording, then listen again while reading your notes and making necessary corrections. This combination strongly reinforces the material.

Always try to preview the material that will be covered before going to the lesson. This is as simple as lightly reading the corresponding sections in the textbook. You may not understand all that you read, but it will sound familiar and be easier to comprehend as your teacher covers it in class. This preview also helps you identify new vocabulary words.

While your instructor is lecturing, don't hesitate to raise your hand to ask a question or get clarification. Many students are shy and

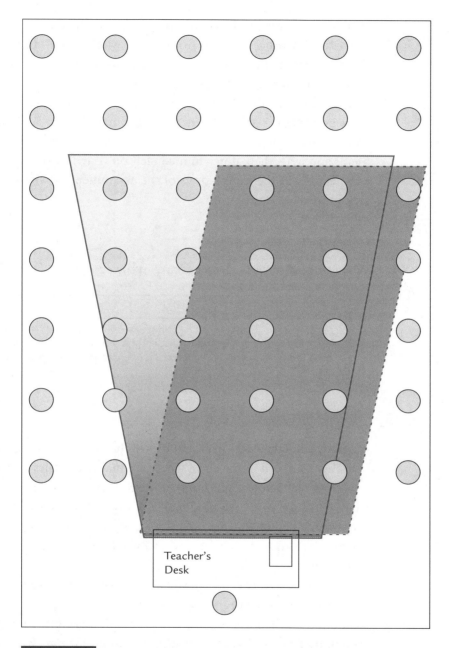

Teacher's
Desk

FIGURE 1.2 **Main areas of focus for a teacher.** The larger shaded area shows where the teacher looks most often. The smaller shaded area shows how that focal area shifts to the right as the teacher uses equipment positioned to the right.

reluctant to speak in class – you may be doing them a favour! Avoid discussing personal issues in front of the whole class – that is better done alone with the teacher, outside the classroom.

Note your teacher's gestures, facial expressions, and voice tone for clues about what they find most important. That material is likely to show up on a quiz or test. Write down any material that is particularly emphasised, or highlight it in your notes. Listen carefully for assignments and write them down immediately on your To Do list. If you are not clear about the expectations of the assignment or when it is due, seek immediate clarification.

LOOK OUT

Why is it best to sit at the front and in the centre of a class?

_____

_____

Answer: You will be more engaged in the class, have the best view and fewer distractions, and be within your teacher's focal area.

Taking Notes

Anybody can take notes in class, but will the notes be good enough to help them succeed in the class? There are many strategies and models for how to take notes, and none of them is necessarily the best. Find what works for you, then use it consistently. Let's review one easy-to-use system (**Figure 1.3**).

Start with an A4 notebook that you will use just for this class. Take your notes on only the front side of the paper and leave about a 5cm margin on the left. The margin will be used for marking key words and concepts later. At the beginning of the lesson, date the top of the page so you know when the material was covered. During the lecture, use an outline format to get as much information down as you can. Use the main concept as a major heading, then indent the information discussed on that topic. When that section ends, either draw a horizontal line to mark its end or leave a couple of blank lines. Don't try to write every word – just the main ideas – and put them in your own words.

RECALL:	03/04/06
Note-	I. Note-taking tips
taking	A. Use outline format
	B. Be concise
	C. Get main ideas
Reviewing	II. Reviewing notes
	A. Review after class
	1. Fill in gaps
	2. Clean up
	3. Replay lecture in my mind
	4. Review within 24 hours – fresh in mind.
3 learning	III. Learning styles
styles	A. Visual – reread and add drawings
	B. Tactile – rewrite or type
	C. Auditory – read out loud or record

FIGURE 1.3   Sample of lecture notes using the outline style and leaving room in the left-hand margin.

Instead of writing out every example, give a brief summary or a one- or two-word reminder. Use abbreviations when possible, and develop your own shorthand. You can often drop most of the vowels in a word and still be able to sound it out later when reading it. Write legibly or your efforts will be useless later. Underline new or stressed terms and place a star or an arrow by anything that is emphasised. Be as thorough as you can, but you will need to write very quickly. The teacher will not wait for you to catch up, so speed is essential.

APPLYING THE THEORY

It is important from the start that you are able to write clearly, as the documentation you will use in your profession may be legally binding, e.g., care plans, drug charts, pre-op and post-op care. Care plans need to be written concisely, giving detailed and relevant information only. You need to develop the ability to précis and be discerning. Record keeping needs to be factual, consistent and accurate: written in such a way that the meaning is clear, and it should be written contemporaneously, *not* a long time after the event.

As soon as possible after the lesson, read your notes and improve them as necessary. Add anything that's missing. Make them clearer and cleaner. Put the concepts in your own words. Next, use the left margin to summarise each section – the main concept, subtopics, and key terms. This column will be your 'Recall' column. Once you are sure all of the key ideas are in the left column, you can cover the right side of the page – the meat of your notes – and quiz yourself on the main points listed in the left column. It makes an easy way to review.

But you are not finished – if you are a tactile learner, rewrite your notes in another notebook or type them into your computer.

Using word processing is good practice for your essay writing. It allows you to combine information from the lesson with textbook notes and anything you get from study sessions. It is helpful to store word processed work electronically as this allows additional material to be added and further changes to be made with ease.

LOOK OUT

Copy and back up your work on a compact disc, memory stick or external hard drive – just in case! A useful way is to email it to yourself. That way you get an extra copy somewhere else.

Visual learners might type and reorganise the notes. Auditory learners can read the notes out loud or record them. You can make flash cards from the key points and terms by writing the term on one side of an index card and its definition or use on the other side. You can add drawings. Review your notes as much as you can during the next 24 hours while the lecture is still fresh in your mind.

TIME TO TRY

Look at the sample notes in Figure 1.3. Now practice: take notes using this style while listening to a one-hour TV programme. Capture the conversations and action in words. You can't get every word down, so paraphrase – put it in your own words so the meaning still comes across. When you're finished, assess how you did.

Can you tell who was talking? _____

Do your notes make sense? _____

Did you capture the main ideas? _____

Did you keep up or fall behind? _____

Do you have breaks in your notes to separate the main conversations and action? _____

What can you do better while taking notes in class?

_____

_____

✔ **QUICK CHECK**

What should you do with your notes after class? _____

_____

_____

Answer: Review them within 24 hours, fill in anything missing, clean them up, put them into your own words, add key concepts and terms to the Recall column, add drawings, make flash cards, record them, rewrite or type them.

Looking for Somebody Special **In Practical Classes**

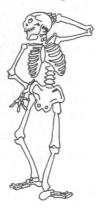

Learning in anatomy and physiology classes doesn't just happen in lectures. Many students put most of their effort into the lecture material and disregard practical classes and the skills lab component. Avoid this. Practicals are the hands-on part of the course, and most people learn better by seeing demonstrations and actually doing the work themselves. Always go to practical classes prepared to take notes and equipped with your textbook if it will be needed.

When in practicals, you may work with a partner or in a group. You will be expected to contribute equally to the team effort, so it is important that you arrive prepared. If you know in advance what the subject will be, read through it and think about what you will be doing. Pay attention to the instructions and especially note any safety precautions. At times you will be working with very expensive equipment and specimens, and perhaps potentially dangerous materials, so always use great care.

Some students try to take shortcuts in practicals so they can leave a bit early. You should value the time to further explore the material covered in lectures. You get to hold the bones, microscopically examine various tissues, use models and charts, and perhaps see real organs or cadavers. It is a unique aspect of your education that reinforces everything else that you are learning. You will spend much time learning structures, and the more you go over them, the better you will recall them. Remember that lectures and practicals are both part of the same class, and try to see how they fit together. Never leave early – there's always more to learn.

APPLYING THE THEORY

Much of the work in your chosen profession will be hands-on practical activities and developing clinical skills. Be aware that you need as much practice as possible before you qualify as you will be working in real-life situations with real people.

✔ **QUICK CHECK**

What are some of the learning advantages gained from attending practical sessions? _____

_____

Answer: Practical classes allow time for exploration, hands-on learning, collaboration and discussion.

 The more time you spend in a practical class or skills lab, the better you will learn. ▪

Your Secret Life **Outside the Classroom**

REALITY CHECK

Answer True or False to each of the following statements:

1. I study the day before a test but rarely study on a daily basis.

   Ⓣ F

2. I mostly review my notes and don't read the textbook.

   Ⓣ F

3. I am too busy to study each day.

   Ⓣ F

4. When I finally get around to it, I study pretty hard for a long time.

   T F⃝

5. I get by fine with cramming.

   Ⓣ F

JUST FOR FUN

Let's see how good a studier you REALLY are! Take a few moments to learn these terms. We will come back to this exercise a bit later.

1. **Frizzled greep**. This is a member of the *Teroplicanis domesticus* family with girdish jugwumps and white frizzles.

2. **Gleendoggled frinlap**. This is a relatively large fernmeiker blib found only in sproingy sugnipers.

3. **Borky-globed dungwinger**. This groobler has gallerific phroonts and is the size of a pygmy wernocked frit.

Stay tuned!

You made it through the day and are ready to head home. Finally! College is done for the day, right? Not if you plan to be successful! The real work begins after class, because most of your learning occurs outside the classroom on your own. This is often the hardest part, for many reasons. We schedule many activities and set aside time for them, but studying tends to get crammed into the cracks. Too often, studying becomes what you do when you 'get around to it'. It is an obligation that often gets crowded out by other daily activities, and the first item dropped from the To Do list.

Too many students only study when they have to – before a test or exam. A successful student studies every day. The goal is to learn the material as you go rather than frantically try to memorise a large amount at the last minute. Here is something you need to know and really take to heart.

You should study for at least two to three hours for every hour spent in class. ■

Simple maths shows you that if you have three lectures on Monday, for example, you should plan to spend from six to nine hours studying that same day!

Schedule Your Study Time

Writing assignments on your To Do list makes them seem more urgent, but that does not cover the daily work that must be done. You must take charge of your time and studying. In addition to specific assignments, each day you should:

- go over that day's notes,

- read the corresponding sections in the textbook,

- test yourself,

- review your notes again, and

- preview the next day's material.

All of this takes time. You must build study time into your schedule or you either will not get around to it or you will put it off until you are too tired to study effectively. The first thing to do is to write your study time into your diary and master calendar, and regard that time as sacred – do not borrow from it to do something else. Be sure to allow break time during study sessions as well – if you study for too long, your brain gets tired and your attention starts to wander, and it takes much longer to do even simple tasks. Plan a 10- to 15-minute break for every hour of studying.

LOOK OUT

If a job seems too large, we put it off, but if we have many small tasks, each alone seems manageable. Break your workload into small chunks. Write them down, partly so you do not forget any, but especially because you will get a great feeling of accomplishment when you complete a task and cross it off your To Do list! Completing a task is also a convenient time to take a mini-break to keep your mind fresh. Many students try to read a whole chapter or cover a few weeks of notes in one sitting. The brain really dislikes that. When studying a large amount of material, divide it into subcategories, then study one until you really understand it before moving to the next.

APPLYING THE THEORY

Try using mnemonics to help you remember a concept. Some of these are acronyms like the one for making student-led learning agreements:
SMART = **S**pecific **M**easurable **A**ttainable **R**ealistic **T**imeframe.

Study Actively

Merely reading your notes or the book is not learning: you must think about the material and become an **active learner**. Constantly ask yourself, 'What is most important in this section?' While reading, take notes or underline key terms and major concepts. Make flash cards. Consider how what you are studying relates to something with which you are already familiar. If you can put the information in a familiar context, you will retain it better. Link what you are learning in theory to what you see in practice.

APPLYING THE THEORY

Watch out in practice for links to the theory you are studying at college. Organise lists of new terms you come across in practice – and make links to the theory in your textbook. Absorbing knowledge can happen in both places.

The best preparation for quizzes and tests is practice. Develop and answer questions as you read. Try to anticipate all the ways your tutors might test you about that material. Recall which specific items your lecturer stressed. Outline the material in each section and be sure to understand how the different concepts are related. Check yourself on the meanings of the key terms. Say the key words out loud and look carefully at them. Do they remind you of anything? Have you heard them before: at home, at work, at college?

Move Past Memorising

This is one of the hardest study traps to avoid. In anatomy and physiology, it may at times seem like there is so much to learn and so little time. Most students at first attempt to just memorise. If you only read your notes and the book, you are using this approach without realising it.

At the beginning of this section, I gave you three items to learn. Without turning back, write down the three names I asked you to learn a few pages ago:

1. _____

2. _____

3. _____

Did you remember them? Now, also without looking back, can you explain each of them to me? _____
(I am betting not)

These three 'things' are fictitious, but my point is that you may, indeed, have memorised the names – it doesn't take much to learn parrot-fashion – but it takes a lot more to understand, especially if the words are unfamiliar, as they often are in your course. If you find that you study hard but the wording of the quiz or test confuses you, I can almost guarantee that you are memorising. The question is worded a bit differently than what you memorised, so you don't realise that you know the answer. You must get past memorising by looking for relationships between the concepts and terms, and really strive for full understanding. Reading often produces memorisation. Active studying produces understanding.

APPLYING THE THEORY

Look for connections – how things are related to one another. The secret to understanding anatomy and physiology is connecting the whole of the body together. If you are anaemic, you don't just have a low red blood cell count, you are also very tired. Think: Why? What? and How? to link it all together.

The Concept Map

A very useful technique for learning relationships is drawing a **concept map**. This is somewhat like brainstorming and similar to a Mind Map. Here is the general process:

1. Start with a blank piece of (preferably) unlined paper.

2. Near the centre, draw a circle and, inside it, list the main concept you will explore.

3. Around that circle, and allowing some space, draw more circles and list in each anything that pops into your mind related to your main concept. Do this quickly and don't think about the relationships yet. Just get your ideas down.

4. Once you've added all your secondary concepts, look at them and think about how they are related, not just to the main concept but to each other as well.

5. As relationships occur to you, draw arrows connecting related concepts and add a brief description of the relationship between each of the concepts.

6. Examine the relationships and you will start to understand how these concepts fit together.

TIME TO TRY

Construct a concept map around the main concept of *energy* by adding arrows to show relationships between the following concepts:

■ cell activity,

■ food,

■ plants,

■ the Sun, and

■ work.

When you are finished, look at **Figure 1.4**, below.

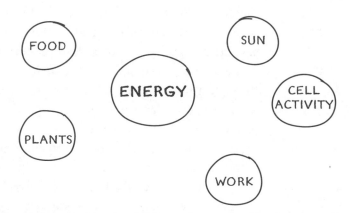

First draw a circle or 'node' for each concept, keeping the main concept, if there is one, near the middle.

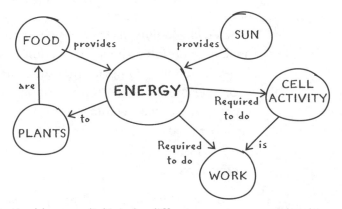

Next, add arrows linking the different concepts to each other, then add brief descriptions of how they are linked. In this example, the arrows show that the sun provides energy to the plants, plants are food that provides energy that is required to do cell activity, which is a type of work.

FIGURE 1.4 Drawing a concept map.

Review

OK, you've been at this study thing for a while and you think you're starting to understand the material. You did all of the above; can you stop now? Almost, but once you think you have the material under control, review it one more time. Repetition is the key to your long-term memory – the more you go over the material, the longer it will stay with you. I always recommend a minimum of three passes, even for easy stuff – read your notes, read the text, re-read your notes – and that assumes you understand the material. Always slow down and go through it more thoroughly if you are struggling with a certain section. Use **active learning** with each pass, then finish with one more review. If you are alert enough to review material right before bed, once you are asleep your brain often continues going over the material without bothering you too much (although one of my students reported a dream in which she was chased by a herd of giant bones!).

The website for your textbook also provides a good way to review subject matter. The address is in your book. Most book websites offer a wide variety of activity options, perhaps including online animations, flash cards, puzzles, objectives, vocabulary lists and quizzes. When taking online quizzes, be sure to do so without looking in your book or your notes – after all, you don't use them for quizzes in class. Not using them provides a better simulation of the classroom experience, and if you do well on the online quizzes without using your notes, you will have confidence in the classroom knowing you have already passed one quiz!

TIME TO TRY

Go to the *Get Ready for A&P for Nursing and Healthcare* website associated with this book, **www.pearsoned.co.uk/getready**

Enter the site and read the Welcome note. Click on the drop-down box and click on Chapter 1: Study Skills. Then press Go.

Determine your starting point by the <u>Pre-test</u> quiz.

Can you relate this to what you have just read in your textbook?

No Cramming Allowed!

I have a very busy life, so my house occasionally gets a bit cluttered. If visitors drop by and it is a 'bad house day', I might quickly grab some of the clutter and cram it into a spare cupboard. After the guests leave, I open the cupboard door to pull out a quilt. What happens? _____

_____

Now imagine what you do to your brain when you cram for an exam. You are essentially opening the cupboard door and cramming stuff in, then slamming the door. When you are taking the test, you open the door to pull out the answer you need, but anything might tumble onto your paper. Cramming at best allows partial memorisation. At worst, it causes the information to get mixed up and you fail. It is a desperate act of superficial studying guaranteed to NOT get you through anatomy and physiology. If you study on a daily basis instead of doing a panicky cram session before a test, you will be calmly reviewing what you already learned well and smiling at the crammers in class.

No Vampires Allowed!

Do you think you can pull an all-nighter and really do well? _____
What do you think are some of the reasons this will not work? _____

_____

If you normally live your life by day, you cannot suddenly override your natural biological clock and expect your brain to stay alert and focused when it knows it is supposed to be asleep. Caffeine may help keep your eyes open, but you'll only be a bit more alert and jittery while still yawning and mentally drifting away from the task at hand. The only reason you would need to stay up all night is if you haven't been studying all along and this is your last option. It is not effective. You will not be mentally alert. You will not be able to focus or think through the material. Your eyes may skim the pages, but you'll struggle to comprehend the words and you'll retain only a tiny amount of the very little that you absorb.

An all-nighter is basically a marathon cram session held at the worst possible time. It simultaneously robs your brain and body of what they need – restoration before the next day. You may be able to stay awake all night, but if you doze off you may oversleep and miss your exam. Or, if you do arrive (I hope you weren't driving with no sleep!), you may get part of the way into the test only to have your brain freeze up on you. If you are prone to 'test anxiety', your defences will be down and you will quite likely freeze and fail. Ah, if only you had been studying all along . . .

For your brain to be kind to you, you have to be kind to its home. You must take care of yourself physically – eat, sleep, exercise and RELAX. ■

✔ **QUICK CHECK**

Why should you study every day if the test is not for two weeks?

_____

_____

Answer: Studying on a regular basis breaks the material into smaller, more manageable pieces that you can master; the material is fresh in your mind, and you will only need to review it before the test.

Strength in Numbers: **The Study Group**

One of the best ways to learn anything is to teach it to someone else, so form a study group or discuss the material with others around you (**Figure 1.5**). Although this may not be the best option for everyone, it is highly effective for many students. As soon as possible, start asking your classmates who wants to be in a study group – you _will_ get people to join. You can quiz each other, discuss the material, help each other and, importantly, support each other. If you study solo, you may not be aware of your weaknesses. Your study partners can help identify them and help you overcome them. A good way to work in a study group is to split up the material and assign different sections to different members, who then master the material and teach it to the group. Each member should also be studying it all on their own – that ensures

FIGURE 1.5   Study groups can be very helpful for staying motivated and focused on course material.

better effort from everyone, and allows other members to correct any errors in a presentation.

Scheduling joint study sessions can be challenging. Many students find that scheduling group sessions before or after class works best. You may want to establish some ground rules, including agreeing to use the time for studying and not for gossiping or just socialising. And although it may be tempting to meet over a pizza, you do want a quiet location where you can freely discuss the material with few distractions. Check with your tutors to see if there are any places that might be available for this.

APPLYING THE THEORY

Communication skills are key to being successful in your chosen profession. This does not just apply to your writing but also to developing the ability to give clear explanations to your clients/ patients about their care. You can practise these skills with your peers.

SQRHuh? **How to Read a Textbook**

The name may sound odd, but **SQR3** is an effective method for studying your textbook. Science textbooks do not read like novels, so you need to approach them differently. This method also works well for reviewing your notes. It stands for:

- Survey

- Question

- Read

- Recite

- Review

During the **survey phase**, read the chapter title, the chapter introduction, any other items at the beginning of the chapter, and all of the headings. This gives you the road map of where you will be going in the chapter. As you skim the chapter, also read all items in bold or italic. Next, read the chapter summary at the end of the chapter.

During the **question phase**, look at the heading of each section and form as many questions as you can that you think may be covered in that section. Write them down. Try to be comprehensive in this step. Ask What? Why? and How? as you read the chapter. By having these questions in mind, you will automatically search for answers as you read and develop critical thinking skills.

Now **read** the chapter for details. Take your time. Adjust your reading speed with the difficulty of the material. Also, keep in mind the questions you developed and try to answer them. Look for the main ideas and how the chapter is organised, linking to this the tables and diagrams.

The next phase is to **recite**. You are working on your ability to recall information. After reading each section, think about your questions and try to answer them from recall. If you cannot, reread the section and try again. Continue this cycle until you can recite the answers: it can help to do this out loud.

Finally, you want to **review**. This helps reinforce your memory. After you complete the previous steps for the sections you're studying, go back to each heading and see if you can still answer all of your questions. Repeat the recite phase until you can. When you have finished, be sure you can also answer the questions at the end of the chapter.

✔ **QUICK CHECK**

What is SQR3 and what do the letters mean?

_____

_____

Answer: It is a system for reading a textbook, and the letters mean **S**urvey, **Q**uestion, **R**ead, **R**ecite and **R**eview.

APPLYING THE THEORY

Always link what you are reading to what you see in real-life situations.

A Place to Call My Own: **The Study Environment**

Briefly describe the location where you plan to do most of your studying.

_____

_____

_____

Now you know how to study effectively, but we often overlook *where* to study. Your options may be limited, so you need to make the best of what you have. Ideally your study spot will be isolated and free of distractions like TV, music and people. At the least, you should minimise distractions.

Do you study in front of a TV that is on? Even if you try hard to ignore it, you will be drawn to it, especially if the material you are studying is tough. Music can be tricky – songs that you know, especially catchy ones, may get you tapping and singing along with them while you think your mind is actively engaged in learning. Soft or classical music can keep you calm and more focused, unless you really dislike it.

Thinking about the study site you listed above, what distractions might you face? _____

_____

How can you minimise them? _____
If you cannot, seek another study spot. _____

For studying, you really need a space that is your own. A desk is a good place (unless it is also the computer desk at which you spend hours playing computer games for fun!). Ideally it will be a place where you do nothing but study, so that when you are seated there you know exactly what your purpose is. If you are having trouble focusing on the task in hand in your study area, get up and walk away briefly. The mental and physical break may help you 'come back' to work, and you won't begin associating this place with struggling. Your study area should be quiet and it should have good lighting to avoid eye strain, a comfortable chair, good ventilation and temperature, and a work surface on which you can spread out.

You will spend a lot of time studying here, so take time initially to set up your study space. The area should be uncluttered and well organised. It should also be inspiring and motivational. Perhaps frame a sign that says '*I WILL be a _____(your career)_____ by _____(your goal date)_____*'. Try the same approach for the goals you set at the beginning of this chapter, and display them boldly and prominently. Consider displaying a photo of your hero, or of someone in the family whom you want to make proud. Parents, you might display a photo of your children with a caption saying something like 'You are my reason', or 'I will teach by example'. Realise that you are their role model – your children's attitudes toward education will be formed by what you do now. With these treasures surrounding you, you're a mere glance away from being reinvigorated if a study session starts to fizzle.

LOOK OUT

Most of your learning is done outside of the classroom. The more efficient your studying, the better you will learn. Your study spot affects your attitude and concentration. The more seriously you take your study location, the more seriously you will study there.

If you live with family or a roommate, you absolutely must stress to them the importance of respecting your study time and study space. Be sure they know your career goals and why they are important to you, and ask them to help by giving you the time and space you need to succeed. Ask them not to disturb you when you are in your study space. If you have small children who want time with their mother or father while you are studying, assuming they have adequate supervision, try getting them to play or study on their own until 'the clock hands are in these positions', then do something fun with them at that time. They will learn to anticipate your time together and to leave you alone if the reward is worthwhile. If you have too many distractions at home, the solution is to study elsewhere. Whether on campus, in the local library, or at a friend's house, you need a distraction-free setting, and if you can't get it at home, remove yourself instead of trying to cope with a poor study space.

✔ **QUICK CHECK**

What are some of the main considerations in selecting your study area? _____

_____

Answer: Few distractions, own space just for studying, comfortable, good lighting/ventilation, sufficient work space, and welcoming.

My, How **Time** Flies!

You know you need to study and that it takes a lot of time, but how will you fit it in? Let's discuss a few ways to budget time for studying. First, be consistent. Consider your schedule to see if you can study at the same time each day. Studying will become a habit more easily if you always do it at the same time. Some students adhere to one schedule on weekdays and a different one on weekends. When scheduling study time, consider your other obligations and how distracted you might be by other people's activities at those times. Don't overlook free hours you might have while on campus. Head to the library, study room, or a quiet corner. This is the ideal time to preview for the next lesson or to review what has just been covered.

TIME TO TRY

This is a two-part exercise designed to help you find your study time.

Part A: Each week has a total of 168 hours. How do you spend *yours?* **Table 1.4** on page 39 allows you to quickly approximate how you spend your time each week.

1. Complete the assessment in Table 1.4 to see how many hours are left each week for you to study.

2. Enter that number here: _____ hours

Part B: Next, turn your attention to **Table 1.5** on page 40.

1. Enter your college timetable, work schedule, and any other activities in which you regularly participate.

2. Now look for times when you can schedule study sessions and write them in.

3. Are you able to schedule 2 to 3 hours of study time per hour of class time? _____

It can be difficult, but it is essential to make the time. Writing it into your schedule makes it more likely to happen.

LOOK OUT

The most efficient use of your study time is in half-hourly segments: most people cannot concentrate on one topic for more than 20 minutes. If you make a timetable for yourself you are more likely to stick to a regular study routine.

Don't overbook! Be sure to build in break time during and between your study sessions, especially the longer ones. Allow for flexibility – realise that unexpected events occur, so be sure you have some extra time available. Also, be sure you plan for and schedule recreation, too. You cannot and should not study all the time, but these other activities do take time and need to be in your schedule as well, so that you do not double-book yourself.

TABLE 1.4 Assessing how your time is spent. For each item in this inventory, really think before answering and be as honest as possible. Items that are done each day must be multiplied by seven to get your weekly total. One item may be done any number of times a week, so you'll need to multiply that item by the number of times each week you do it. After you have responded to all the questions, you'll have an opportunity to see how many hours remain during the week for studying.

Where does your time go? Record the number of hours you spend:	How many hours per day?	How many days per week?	Total hours per week: (hours × days)
1. **Grooming**, including showering, shaving, dressing, make-up, and so on.			
2. **Dining**, including preparing food, eating, and cleaning up.			
3. **Commuting** to and from college and work, from door to door.			
4. **Working** at your place of employment.			
5. **Attending college**.			
6. **Doing chores** at home, including housework, mowing, laundry, and so on.			
7. **Caring** for family, a loved one, or a pet.			
8. On **extracurricular activities** such as clubs, church, volunteering.			
9. **Doing errands**.			
10. On **solo recreation**, including TV, reading, games, working out and so on.			
11. **Socialising**, including parties, phone calls, chatting with friends, dating and so on.			
12. **Sleeping** (don't forget those naps!).			
Now add all numbers in the far column to get the total time you spend on all these activities.			
		Hours/week	168
		Total hours spent on other activities	—
		Left for studying =	

TABLE 1.5   My study schedule.

Time	Monday	Tuesday	Wednesday	Thursday	Friday	Saturday
6:00 AM						
7:00 AM						
8:00 AM						
9:00 AM						
10:00 AM						
11:00 AM						
Noon						
1:00 PM						
2:00 PM						
3:00 PM						
4:00 PM						
5:00 PM						
6:00 PM						
7:00 PM						
8:00 PM						
9:00 PM						
10:00 PM						
11:00 PM						
Midnight						

APPLYING THE THEORY

Healthcare practitioners need to be good at managing time so they can prioritise workloads on the wards or caseloads in the community.

Putting It to **The Test**

If I had a pound for every student who said they have test anxiety . . .

Some people really do suffer from true test anxiety, but the majority of students who claim to have this condition believe it to be true not because of an actual diagnosis, but rather because they get very nervous and may go blank during tests. If I ask a class who among them suffers from test anxiety, most hands go up. By the end of the course, with some coaching, the number is far less. Why? They have learned how to take tests and how to stay calm. If you do suffer from true test anxiety, consult with your personal tutor right away so they can put you in contact with the support services you need to understand your condition and learn how to conquer it.

Most people dread taking tests and experience some anxiety when taking them. Not surprisingly, the better prepared you are for an exam, the less worried you will be. The best remedy for the stress you associate with taking tests is to be very well prepared. If you know you understand the material, what is there left to worry about?

Some people get very anxious before tests because they fear they will not do well. This may be because they know they are not prepared. Again, the remedy is simple: study well. Anxiety can also arise from a bad past experience. If you have done poorly on tests in the past, your self-confidence may be lacking, so you anticipate doing poorly. That may lead to cramming and memorising instead of truly learning, and may cause you to become excessively nervous during the test, which can cause poor performance. All you need is a couple of good grades on tests to get your confidence back!

If you are a nervous test taker, don't study for about an hour immediately before your exam. Students who complain of test anxiety are frantically reviewing their notes right up to the moment they go in to the test. They have been trying to quickly glance back over everything while racing against the clock. No wonder they are stressed! Remember that your brain needs time to process the information. When you cram information into the 'cupboard', who knows what will fall out when you open the door during the test.

If you have studied well in advance and don't get very nervous at exam time, you might want to glance quickly through your notes beforehand, but only if you have time to do so and still allow *at least* a half-hour to relax and mentally prepare for your test. The half-hour off allows your brain to process the information while you relax. Try getting a light snack so you are alert – a heavy meal could make you drowsy during the test. Walk around to release nervous energy. Listen to music that makes you happy. Sit comfortably, close your eyes, and breathe deeply and slowly while you picture yourself in a very relaxing setting – maybe on a tropical beach, curled up on your couch with a good book, or out on a boat fishing. Focus on how relaxed you feel and try to hold that feeling. Now, staying in that mood, concentrate on how well you have studied and keep reminding yourself that:

- I have prepared very well for this test.

- I know this material very well and I answered all questions correctly while studying.

- I can and *will* do well on this test.

- I refuse to get nervous over one silly test, especially because I know I am ready.

- I am ready and relaxed. Let's get it done!

Test-taking Tips

There are also strategies you can use while taking the test. Let's see what your current strategies are. Complete the survey in **Table 1.6**, and then we will discuss specific strategies.

During an exam, be careful – read each question *thoroughly* before you answer. This is especially true of multiple choice and true/false questions. We know the answer is there, so our eyes tend to get ahead of our brains. We skim the question and jump down to the answers

TABLE 1.6   **Self-evaluation of test-taking skills.** For each of the following valuable test-taking skills, mark if you do each one always, sometimes, or never. Highlight any that you do not currently use that you think might help you be more successful.

Test-taking skill	Always	Sometimes	Never
1. While studying my notes and the book, I think of and answer possible test questions.			
2. I use online practice quizzes when they are available.			
3. I avoid last-minute cramming to avoid confusing myself.			
4. I scan the whole test before starting to see how long it is and what type of questions it contains.			
5. I do the questions I am sure of first.			
6. I budget my time during a test so I can complete it.			
7. I answer questions with the highest point values first.			
8. I read all the answer options on multiple choice questions before marking my answer.			
9. I know what key words to look for in a multiple choice question.			
10. I use the process of elimination during multiple choice or matching tests.			
11. I know what key words to look for in essay questions.			
12. I look for key words like *always*, *never* and *sometimes*.			
13. When I am unsure of an answer, I go with my first answer and fight the urge to change it later.			
14. I try to answer everything even if I am uncertain, instead of leaving some questions blank.			
15. I check my answers before turning in a test.			

before even trying to mentally answer the question. Slow down and think before moving to the answers. Otherwise you may grab an answer that sounds familiar but is incorrect. If you have trouble keeping your eyes off the answers, cover them with your hand or a ruler until you finish reading the question and think of the answer on your own. Then reveal the answers one by one to decide which is correct.

If you do not know the answer initially, take a deep breath and think of all you do know about the words in the question. Often this is all you need to recall the answer. This is when those concept maps you made will really help you.

Use the process of elimination. If you are not sure which answer is correct, can you eliminate any you know are incorrect? Narrow down your choices. Don't make a guess unless the process of elimination fails you; however, guessing is usually better than leaving a question unanswered, unless you lose points for wrong answers. On short-answer, fill-in-the-blank questions, and essays, always write some-thing.

After you answer a question, read your answer to be sure it says what you want it to, then leave it alone. Once you move on, resist the temptation to go back and change your answers, even those of which you were unsure. Often we have a gut instinct to write the correct an-swer: perhaps we are recalling it at some subconscious level, but the very act of going back is a conscious reminder of uncertainty, and we often choose something different only because we doubt ourselves.

When answering multiple choice or true/false questions, ignore any advice that suggests you should select one answer consistently over others. Also, don't worry if you choose the same answer several times in a row, thinking the teacher would not structure a test that way. I can't speak for all teachers, but I do not personally know any who give much thought to the pattern the answers will make on the answer sheet, so neither should you.

Here are a few more pointers:

■ Note the wording on questions. Key words to look for that can change an answer are *always, sometimes, never, most, some, all, none, is* and *is not.*

■ Glance over the exam as soon as you receive it, so you know what to expect, then budget your time accordingly.

■ Look for questions on the backs of pages so you don't miss them.

- Tackle easy questions first. They may provide hints to the tougher ones.

- Be aware of point values and be sure the questions with the greatest point values are done well. Often essay questions – which are usually worth more points – are at the end, and some students run out of time before reaching them, losing significant points and seriously hurting their mark.

- If you have trouble writing essay answers, recall all you know about the topic, organise in your mind how you would explain it to someone, then write down your thoughts as if you are writing yourself a letter about what to say.

- If a question has multiple parts, be sure to answer each part. This is especially true for essays.

- If you are asked for a *definition*, give a book explanation of what the term or concept means. If you are asked for an *example*, list an example and explain why it is an example of the concept. If you are asked to explain a concept or term, approach it as if you are trying to teach it to a 6-year-old. Assume the reader has no prior knowledge.

- Be very thorough and specific in your answers. The marker cannot get inside your head to decide if you knew it or not, so your words must very literally convey your meaning.

When a test is returned, record your grade. Be sure to review the test to see which questions you got wrong and why, and then make notes to go back and review that material. Remember – it may come back to haunt you on a bigger test or on the final exam, and you should know it anyway.

LOOK OUT

Remember you are not learning this to pass a test, but so that you are better able to carry out your chosen profession.

✔ **QUICK CHECK**

How can you slow yourself down when taking a multiple choice or true/false test? _____

_____

Answer: Cover the answers with your hand while you read the question, and don't look at them until you think of the answer.

TIME TO TRY

Go to *Get Ready for A&P for Nursing and Healthcare* on your computer.

Enter the web address **www.pearsoned.co.uk/getready**

Go to Chapter 1: Study skills.

Find out what you learned by doing the <u>Post-test</u> quiz.

Look at the <u>Glossary</u> for terms that you may have learned during the study of this chapter.

Through the Looking Glass: **Individual Accountability**

I hope you now have some insight into the learning process and have developed new strategies to improve your success, not just in anatomy and physiology, but in all of your classes. One more area needs to be discussed, though, and that is your responsibility and attitude. When we get frustrated, we often look elsewhere for the cause, even when it may be right in front of us. I have watched poorly prepared students transform themselves into really good students and I have seen good students drop out as they start getting really bad grades. Many factors can contribute to these changes, but a common thread is always attitude and accountability. Here are three facts you need to firmly implant in your mind:

1. *You*, and nobody else, chose to pursue this academic and career path.

2. *You*, and nobody else, are responsible for attaining the success you desire.

3. *You*, and nobody else, earn the grades you get.

You must do everything you can to guarantee your success – nobody will do it for you. That means always accepting responsibility for your own effort. No excuses. To stay on track, you must know exactly what you want and always stay focused on where you are going. At times, you may not feel that you can keep up, but instead of giving up or slacking off, you need to refocus on where you are going and why it matters to you. Always set short-term and long-term goals. Write them down and put them up where you will see them often. You are responsible for keeping yourself motivated. Learn to visualise your success – see yourself in your future career. Think about how your life will be. Dream big, then go after that dream with all you have.

An important part of any journey is to anticipate roadblocks before you hit them. Think carefully about any possible obstacles to your success, then plan around them. You have an unreliable car? Find someone you can ride with in emergencies. You have small children at home? Have daycare lined up and a back-up plan for when your child is too ill to go to daycare. You have a learning disability? Immediately contact student support services or your personal tutor to find out what services are available to assist you. Make a list of anything that might get between you and success, then write down at least two possible solutions so you have a main strategy and a back-up plan.

Finally, consider those around you. Family and friends must know your goals and understand how important they are. But do they support you? I have had women whose husbands burned their books because they felt threatened that their wives might no longer need them once a degree was attained. And there are certainly more subtle means of sabotage. Perhaps your friends needle you because you don't go out as much, or they say you're no fun anymore. Your significant other complains that you don't get enough time together. Relatives accuse you of thinking you are better than them because you are getting some education. Realise that when someone changes, whether through education or something else, those who know them may feel excluded, threatened, left behind, even envious. You can try to assure them how much you still value and need them in your life, but don't let them distract you from your mission.

You must surround yourself with supportive people who are happy and proud of you, who celebrate your victories, and who want for you what you want. They will help you succeed. It may be your study group or others around you who will help you study. Perhaps they will watch your children so you get some quiet time. On the other hand, anyone who ridicules you or is upset by your new schedule and goals is really not a friend you want around. Make new friends in class who share your goals and guard yourself from those who would derail you. Stay on track, and you will soon be living your dream. Good luck!

FINAL STRETCH!

Now that you have finished reading this chapter, it is time to stretch your brain a bit and check how much you have learned.

RUNNING WORDS

At the end of each chapter, be sure you have learned the language. Here are the terms introduced in this chapter with which you should be familiar. Write them in a notebook and define them in your own words, then go back through the chapter to check your meaning, correcting as needed. Also try to list examples when appropriate.

Visual learner	Concept map
Auditory learner	Active learning
Tactile learner	SQR3
Active learner	

TIME TO TRY

Now look at the online Glossary in the *Get Ready for A&P for Nursing and Healthcare* website. Make a list of your new vocabulary and check out the meanings. Making a list like this is a good skill to develop for all aspects of your work.

WHAT DID YOU LEARN?

In the left-hand column below, write your approach before reading this chapter. In the right-hand column, list any changes you plan to make to ensure your success in this class.

What I have done before this chapter	What I will do to improve

During lectures:

Note-taking:

Study habits:

Textbook reading:

My study place:

Time management:

Test-taking:

List the three areas in which you think your study skills are the weakest, and ways in which you plan to improve them.

1. _____

2. _____

3. _____

APPLYING THE THEORY

Academic writing skills and critical thinking skills are required in order to become a professional and use an evidence base to convey your work to others.

WEB RESOURCES

Here are some additional online resources for you:

■ *Study Skills Website of the University of Surrey*

http://www.surrey.ac.uk/Skills/pack/contents.html

Try clicking on the <u>Communications</u> link and look at the <u>Gathering Information</u> link underlined at the bottom of the page.

■ *Skills4Study*

http://www.palgrave.com/skills4study

Look at this free website – especially at the <u>Study Skills</u> icon in green on the Home page. There are also MP3 audio downloads about <u>Tricks of the Writer's Trade</u>, <u>Exam skills</u> and <u>Presentation skills</u>.

■ *INTUTE*

www.intute.ac.uk

The Health and Life sciences pages of Intute are a free online service providing you with access to the very best web resources for education and research, evaluated and selected by a network of subject specialists.

There is a free Internet tutorial to learn to find information resources in allied health disciplines for your online research. Click on <u>Virtual Training Suite</u>.

■ *BBC Education Study Skills website*

http://www.bbc.co.uk/education/asguru/studyskills/

Select a section from the drop-down box. Choices include: <u>What style of learner are you</u>?, <u>Effective note-taking</u> and <u>Revision skills</u>.

■ *Cook Counseling Center Study Skills Inventory*

http://www.ucc.vt.edu/studyskills/aassaform.htm

The purpose of this inventory is to find out about your own study habits, attitudes and skills.

■ *Concept Mapping*

http://www.cet.edu/ete/pbl2.html

Try mindmapping for free at:

http://www.conceptdraw.com/en/resources/ mindmap/main.php

2 Basic Maths Review

Crunching the Numbers

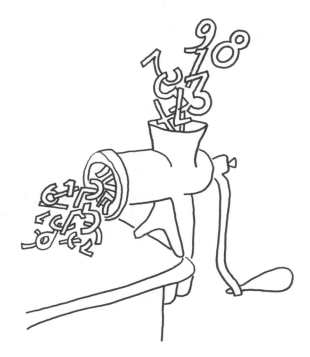

When you have completed this chapter, you should be able to:

■ Solve maths problems involving multiple operations, fractions, decimals and percentages.

■ Calculate the mathematical mean and explain its importance to physiology.

■ Work with exponents, numbers in scientific notation, ratios and proportions.

■ Understand the units of the metric system and be able to convert between units accurately in clinical practice.

■ Perform basic measurements in a clinical setting.

■ Read tables, graphs and charts.

YOUR STARTING POINT

Answer the following questions to assess your maths skills.

1. $2/3 \times 3/4 =$ _____

2. Express 50% as a fraction: _____ As a decimal:
   _____

3. What is the *mean* of 27, 33 and 36? _____

4. In scientific notation, 1000 = 10 _____

5. 30% of 200 = _____

6. Which is longer, 1 yard or 1 metre? _____

7. In the metric system, the basic unit of volume is the _____.

8. Assume you take 15 quizzes by the end of the course and get
   9 As and 6 Bs. Express this as a ratio. _____

9. Is a triple beam balance used to measure volume, mass or
   distance? _____

10. On a graph, the vertical line is the _____ axis.

How Much Wood Would a Woodchuck Chuck?
Maths in Science

You probably remember doing story problems when learning maths in your younger years. Those problems helped you see how maths can be used. Many students are surprised to learn that they have to use maths in anatomy and physiology, but you must remember that science – all science – deals with that which is testable. A scientific test, as you know, is called an experiment. Results collected from experiments are called **data** and, more often than not, the data are numbers. When you try to make sense of the data, you are working with numbers, and that means maths.

Answers: 1. 1/2, 2. 1/2, 0.5, 3. 32, 4. 10³, 5. 60, 6. 1 metre, 7. litre, 8. 9:6 = 3:2, 9. mass, 10. y.

You will probably do some experiments in class and then analyse the data. In addition, many aspects of body function have 'normal' conditions that are often expressed in numerical values. For example, normal body temperature is 37°C, normal blood pressure is 120/80 mmHg, and a normal pulse is around 70 to 80 beats per minute. You will calculate various physiological values. For example, the amount of blood that leaves the heart each minute is called cardiac output, which is calculated by multiplying the number of heartbeats per minute (heart rate) by the amount of blood leaving the heart with each beat (stroke volume). You will work with chemical solutions and you will need to understand their concentrations. You will measure in metric units, refer to percentages and ratios, and interpret graphs and charts.

Many students entering this class need to dust off their maths skills a bit. Some of you may only need a brief reminder of what you learned before, others may need to learn it again. Regardless of your maths history, a quick refresher will help you to better understand the numbers.

You need to have sufficient arithmetic skills to safely calculate drug doses, monitor, record and interpret patients' data.

From the Beginning: **Basic Maths Operations**

You might need to do some complicated computations in class, so it is good to remember the basic rules. Let's zip through them for a quick refresher. I assume you can add and subtract, so we'll skip those operations (see web resources if you need help with addition and subtraction).

Multiplication   Multiplication problems can be done in any order: $3 \times 4 \times 2 = 24$, or $2 \times 3 \times 4 = 24$. The answer to the equation is called the **product**. Recall that multiplying any number by 1 does not change the number, while multiplying any number or numbers by 0 gives you 0. If you have no money and you multiply it by 3, you still have no money!

In more complicated equations, an operation may be set off in parentheses or brackets. If a number appears immediately to the left of a parenthesis or bracket, multiply by that number even though there is no multiplication sign. Don't forget to work out what is inside the brackets first.

$$3(6 - 2) = 3 \times (6 - 2) = 3 \times 4 = 12$$

Multiplication problems are sometimes represented with **exponents**. Consider 2^4, which is read as 'two to the power of four', and 10^3, which is read as 'ten to the power of three'. These examples are really just a shorthand way of expressing these multiplication problems:

$$2^4 = 2 \times 2 \times 2 \times 2 = 16 \qquad 10^3 = 10 \times 10 \times 10 = 1000$$

Division   Like subtraction, division problems must be done from left to right. Consider this example:

$$10 \div 5 = 2,$$
$$\text{but } 5 \div 10 = 1/2$$

Read it out aloud – 'ten divided by five'. This will help you to identify the number to be divided. In division, the number being divided is called the **dividend**, the number by which it is divided is the **divisor**, and the total is the **quotient**. In our first example, 10 is the dividend, 5 is the divisor, and 2 is the quotient.

Multiple Operations   I am sure you remember all of this, but let's try some more complicated problems that involve more than one mathematical operation. Parentheses or brackets are often used in equations when there are multiple operations. Think of them as directors telling you how to proceed. Always do operations within these structures first. Let's see why this matters.

$$8 - (2 \times 3) = 8 - (6) = 2,$$
$$\text{but } 8 - 2 \times 3, \text{ done in that order, } = 6 \times 3 = 18$$
$$\text{OOPS!}$$

LOOK OUT

Some drug calculations involve using the patient's weight or surface area in order to give a very accurate dose.

Always approach an equation by first doing any operations within parentheses. ■

Here are some simple rules to help ensure that you perform mathematical operations in the correct order.

1. First, do all operations inside the parentheses or brackets.

2. Next, multiply out any exponents.

3. Then do all multiplication and division equations, moving from left to right. Remember that 'of' in a sum means the same as multiply.

4. Finally, do all addition and subtraction problems, again from left to right.

TIME TO TRY

Are you with me so far? Try the following problems.

1. $4 \times (9 - 6) + 10 =$ ___22___

2. $3^3 \div 9 - 4 + 5 =$ _____

3. $6^2 - 2(5 - 2) + 4 - 2 =$ _____

Let's see how you did.

Problem number 1: The correct answer is 22. First do what is in the parentheses (rule 1): $(9 - 6) = 3$, so the problem becomes $4 \times (3) + 10$. Next, do the multiplication (rule 3): $4 \times 3 = 12$, so the problem becomes $12 + 10$. Finally, do the addition (rule 4): $12 + 10 = 22$.

Problem number 2: The correct answer is 4. There are no parentheses, so you start with the exponent (rule 2): $3^3 = 3 \times 3 \times 3 = 27$, and the problem becomes $27 \div 9 - 4$. Next, do the division (rule 3): $27 \div 9 = 3$, so the problem becomes $3 - 4 + 5$. Finally, do the addition and subtraction from left to right (rule 4), and you get $3 - 4 = -1$, then $-1 + 5 = 4$.

Problem number 3: The correct answer is 32. Start in the parentheses (rule 1): $(5 - 2) = 3$, so the problem becomes $6^2 - 2(3) + 4 - 2$. Next, take care of the exponent (rule 2): $6^2 = 6 \times 6 = 36$, so the problem becomes $36 - 2(3) + 4 - 2$. Now, do the multiplication and division from left to right (rule 3): $2(3) = 2 \times 3 = 6$ and $4 - 2 = 2$, so the problem becomes $36 - 6 + 2$. Finally, do the addition and subtraction from left to right (rule 4): $36 - 6 = 30$, then $30 + 2 = 32$.

As you see, some mathematical equations can be long and somewhat complicated, but if you keep the basic rules in mind and tackle them step by step, they become quite manageable.

What Do You **Mean** You Are Normal?

Do you know what 'normal' body temperature is? Sure you do – 37°C. You have likely known that since you were a small child. But what is YOUR normal temperature? Mine is rarely 37°C and yours may not be either. 'Normal' blood pressure is 120/80 mmHg yet mine often runs lower than that. Just when I started thinking I might be abnormal, I reconsidered what the term 'normal' really means.

In physiology, the term **normal** means **average**, and in maths, another term for average is **mean**. We refer to many normal values – temperature, blood pressure, pulse, respiratory rate . . . the list goes on and on. When you see these, realise that they are average values and an individual person may have a different normal value – what is normal for them may not be average for the whole population. To better understand this idea of normal, you need to know how to calculate the average, or mean, value.

Let's say you're doing a practical on the cardiovascular system and you're measuring pulse. You are instructed to do three trials and then calculate the mean pulse. Your three trials give you the following data:

Trial 1: 72 beats per minute

Trial 2: 74 beats per minute

Trial 3: 79 beats per minute

To find the mean of a group of numbers, simply add them all together then divide the total by how many numbers you added. For your data, you would add the three pulse rates, then divide by 3:

$$72 + 74 + 79 = 225 \qquad 225 \div 3 = 75 \text{ beats per minute}$$

Did you notice that the mean is not one of the original numbers? It does not have to be. It is the average of all three numbers.

LOOK OUT

A single measurement of vital signs, such as pulse and blood pressure, is not useful for making a decision about a patient's condition. These parameters vary with stress, exercise, food intake and age as well as illness. The observations need to be taken regularly over a period of time so that trends can be seen.

Here is a tip to help you with means and with most maths problems – learn to predict your results! The mean of a group of numbers will be somewhere between the highest and the lowest of the numbers you are averaging. If your value does not fall in that range, check to see if you made an error. Common mistakes include missing a number during the addition or dividing by the wrong number. If you estimate your result first, you can more easily recognise errors if they occur.

TIME TO TRY

Calculate the mean of these body temperatures:

37.2°C, 36.8°C, 37.9°C and 38.1°C

1. What is the mean? _____ °C

2. What does this mean *mean*? _____

_____

If you did this correctly, you should have a mean of 37.5°C, even though that was not one of the original temperatures listed. So, as stated earlier, the physiological 'normal' value is a mean, and individuals' normal temperatures will vary around that mean.

WHY SHOULD I CARE?

All science is based on data and experimental trials and there is a certain amount of error possible with each trial. Consider the pulse values we used as examples. Three trials gave us three results. Using the mean helps to minimise the error from individual trials.

In physiology, means, or normal values, are important because they give us a reference point for how the body is working. If someone's temperature is 39.6°C, we are pretty sure there is a problem because it is significantly far from the normal value. We use means for comparison – the mean tells us what 'normal' function should be so we can recognise abnormal function.

✔ QUICK CHECK

What is meant by saying that normal human heart rate is 80 beats per minute?

Answer: It means the average, or mean, heart rate for humans is 80 beats per minute.

APPLYING THE THEORY

Q. Mrs Andrews has had an abdominal operation. Following surgery, her pulse rate and blood pressure are recorded. Why should these measurements be carried out frequently?

A. A gradual rise in pulse rate and/or drop in blood pressure may be a sign that Mrs Andrews has a haemorrhage or is suffering from post-operative shock and may need further medical attention.

Meet My Better Half: **Fractions, Decimals and Percentages**

Working with whole numbers is rather easy and is second nature to most of us. Fractions, however, are often a faded, distant memory. Related to fractions are two other ways of expressing values: decimals (or decimal fractions) and percentages. In anatomy and physiology, all three of these will be used. For example, the micrometre (μm), a common unit for measuring the size of microscopic structures, is a tiny fraction of the more familiar millimetre – a micrometre is 1/1000 of a millimetre, to be exact. As you just saw, normal body temperature is reported as 37°C and finally, about 60 per cent of the body of an average adult male is water. You will discover that many physiological values are reported in any of these formats, so you want to be comfortable with their use.

Fractions

A fraction is a part of a whole, whether it is something such as a half (1/2) of an apple or a fraction of a minute, e.g. 15 seconds is 1/4 of a minute, which can also be written as a decimal fraction, 0.25 minutes.

Fractions are written as a/b (sometimes called vulgar fractions), in which a and b are both whole numbers and b is not 0. The first (top) number is called the **numerator**, and the one on the bottom is the **denominator**. A fraction represents parts of some whole group (**Figure 2.1**). For example, 3/4 represents 3 equal parts out of 4 equal

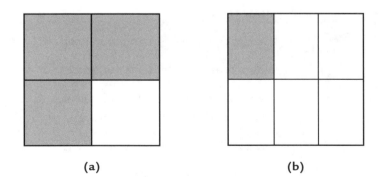

(a)   (b)

FIGURE 2.1   Fractions represent some part of a whole. In each of these squares, the shaded area represents the fraction that is listed below. **(a)** This square is divided into four equal parts, and three of the four are shaded = 3/4. **(b)** This square is divided into six equal parts, and one of the six is shaded = 1/6.

parts, where the 4 equal parts make up the whole (Figure 2.1a). Whole numbers can be represented as fractions as well. The whole number simply becomes the numerator, and the denominator is 1, so 3 = 3/1.

✔ **QUICK CHECK**

For 5/8, what is the numerator? _____ The denominator?
_____ Express 6 as a fraction: _____

Answers: 5 is the numerator, 8 is the denominator, and 6 as a fraction is 6/1.

Reducing Fractions   **Equivalent fractions** have the same value even though they appear to be different. Consider the following fractions:

$$1/3 \qquad 2/6 \qquad 4/12 \qquad 7/21$$

All of these numbers have the same value: 1/3. To see this, you need to **reduce** the other fractions. This is done by finding the **greatest common factor (GCF)** for each fraction. The greatest common factor is the largest whole number that can be divided into both the numerator and the denominator. Consider 2/6. Both the numerator (2) and the denominator (6) are divisible by 2, which is the greatest common factor. If you do the division, you see that 2 ÷ 2 = 1 and 6 ÷ 2 = 3, so 2/6 becomes 1/3.

TIME TO TRY

Look at the other fractions we listed: 4/12 and 7/21.

What is the greatest common factor for 4/12? _____

Divide the numerator by that factor: _____
Divide the denominator by that factor: _____

What is the reduced fraction? _____

What is the greatest common factor for 7/21? _____

Divide the numerator by that factor: _____
Divide the denominator by that factor: _____

What is the reduced fraction? _____

How did you do with Time to Try? You should have found that the greatest common factor for 4/12 is 4, so 4 ÷ 4 = 1 and 12 ÷ 4 = 3. Thus, the fraction 4/12 reduces to 1/3. Similarly, 7/21 has a greatest common factor of 7, and 7 ÷ 7 = 1, and 21 ÷ 7 = 3 so, again, 7/21 reduces to 1/3. After using a number to reduce the fraction, check your result to see if it is in its simplest form or if it can be reduced further.

✔ **QUICK CHECK**

What is the most reduced form of each of the following fractions: 60/90, 25/100, and 18/54?

Answer: 60/90 has a GCF of 30 and reduces to 2/3; 25/100 has a GCF of 25 and reduces to 1/4; 18/54 has a GCF of 18 and reduces to 1/3.

Multiplying and Dividing Fractions When doing mathematical operations with fractions, the rules are different for multiplication and division than they are for addition and subtraction. For multiplication, you simply multiply the numerators in one step, then multiply the denominators. Consider 2/3 × 3/4. The numerators are 2 and 3. Multiply them to get 6, and that goes on top. Next, multiply the two denominators, 3 × 4, to get 12. So the product is 6/12, which reduces to 1/2:

$$\frac{2}{3} \times \frac{3}{4} = \frac{6}{12} = \frac{1}{2}$$

Let's try another: 1/3 × 2/5 × 3/4 = _____

First, multiply all the numerators (1 × 2 × 3 = 6), then multiply all the denominators (3 × 5 × 4 = 60) and you get 6/60, which reduces to 1/10.

To multiply fractions, first multiply all the numerators, then multiply all the denominators. Reduce the result as needed. ▪

Dividing fractions may seem difficult at first, but a simple trick actually makes it easy! These problems may be written two different ways:

$$\frac{4/5}{2/3} \text{ or } 4/5 \div 2/3$$

Solving them is easy. First, invert (flip) the second fraction, which is the divisor: 2/3 becomes 3/2. Then you simply multiply the two fractions:

$$4/5 \div 2/3 = 4/5 \times 3/2 = 12/10$$

Now, 12/10 can be reduced to 6/5. Here the numerator is larger than the denominator, which means this fraction is greater than 1. Usually when this happens, it is best to express the answer as a mixed number – one combining both whole numbers and fractions. To do this, first reduce the fraction: 12/10 = 6/5. Then realise that 6/5 = 5/5 + 1/5. Since 5/5 equals 1, the mixed number would be 1⅕.

To divide one fraction by another, first invert the second fraction to turn it into a multiplication problem. Next, multiply the numerators, then multiply the denominators. Finally, reduce the result. ▨

Adding and Subtracting Fractions   To add or subtract fractions, they must first be in the same format. Adding is straightforward if the fractions have the same denominators (the bottom part of the fractions), when you can simply add the numerators (the top part of the fractions).

$$1/6 + 3/6 = 4/6 \quad \text{which can be reduced to 2/3}$$

This doesn't work for fractions with different denominators, as in the next example.

$$1/2 + 1/4$$

You need to know the *lowest common denominator*. This is the smallest number that can be divided by both denominators.

In the above example, 1/2 and 1/4, the denominators can both divide into 4.

$$1/2 \quad + \quad 1/4$$

$$\frac{2}{4} \quad + \quad \frac{1}{4} \quad = \quad \frac{3}{4}$$

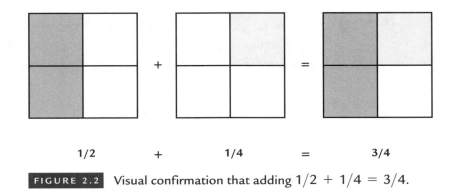

1/2   +   1/4   =   3/4

FIGURE 2.2 Visual confirmation that adding $1/2 + 1/4 = 3/4$.

You have now made equivalent fractions by multiplying the numerator and the denominator of 1/2 by 2 to give 2/4 and the numerator and denominator of 1/4 by 1 to give 1/4. You will notice that now both denominators are 4. You then need only add the numerators.

$$2/4 + 1/4 = 3/4$$

More examples can be found for you to practice on the recommended websites at the end of the chapter.

Figure 2.2 illustrates this for you.

Subtracting Fractions   This process also requires finding the *lowest common denominator* for the fractions and taking away the numerators when you have the fractions in equivalent form.

Look at the following $1/3 - 1/4$

In this example, 1/3 and 1/4 the denominators can both divide into 12.

$$\frac{1/3}{} \qquad - \qquad \frac{1/4}{}$$

$$\frac{4}{12} \quad - \quad \frac{3}{12} \quad = \quad \frac{1}{12}$$

You have now made equivalent fractions by multiplying the numerator and the denominator of 1/3 by 4 to give 4/12 and the numerator and denominator of 1/4 by 3 to give 3/12. You will notice that now both denominators are 12. You then need only subtract the numerators:

$$4/12 - 3/12 = 1/12$$

To add or subtract fractions, use a common denominator to put the fractions in a common form, then add or subtract the numerators only. Remember to always subtract from left to right. ▪

✔ QUICK CHECK

Solve these problems:

1. 3/5 + 1/4 + 1/10 = _____

2. 9/16 − 3/8 = _____

Answers:

1. **19/20.** The lowest common denominator is 20, so the fractions are changed to equivalent fractions. 3/5 = 12/20, 1/4 = 5/20 and 1/10 = 2/20. Add numerators 3 + 5 + 2 = 19 and place over the lowest common denominator, giving the answer 19/20.

2. **3/16.** The lowest common denominator is 16, so the equivalent fractions are 3/8 = 6/16, so 9/16 − 6/16 = 3/16.

Decimals

Decimals are common in science, and anatomy and physiology are no exception. All decimals are based on 10 in a very specific way – each place in the number represents a multiple of 10. The value *increases* 10 times for each space that you move to the left of the decimal, and it *decreases* 10 times for each space to the right.

Let's consider this number: **12345.6789.** You know how to read the numerals to the left of the decimal point. They make 12345. Moving from the decimal to the left, you can see how the spaces represent, in order, 1s, 10s, 100s, 1000s and 10,000s (**Figure 2.3**). Our number represents 10,000 + 2000 + 300 + 40 + 5. The spaces to the right of the decimal represent fractions: tenths, hundredths, thousandths and so on. Thus in our number, as you go to the right of the decimal, the numerals represent the fractions 6/10, 7/100, 8/1000 and 9/10,000.

Converting Decimals   As with fractions, decimals allow you to express a number more precisely than you can with whole numbers. In fact, you can think of fractions as division problems that give you a decimal value, so fractions can be converted to decimals. For example,

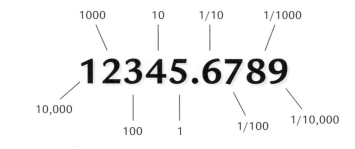

FIGURE 2.3   Each space around a decimal point reflects a change by a factor of 10.

1/2 is 1 ÷ 2, and that equals the decimal 0.5. Note the 0 here – it is used as a place holder so that you know where the decimal point belongs. If you use a calculator to find 1 ÷ 3, the answer will be 0.333333. . . . This is known as a **repeating decimal**. When you have a repeating decimal, one option is to round it off. If the last number is less than 5, round it down; if it is 5 or higher, round it up. For example, 0.33333 . . . would round down to 0.33, and 0.66666 . . . would round up to 0.67. It is usual to round off to two decimal places, as here.

Decimals can also be converted into fractions. The value 0.25 represents 2/10 + 5/100. When you add them, remember to first get a common denominator. You get 20/100 + 5/100 = 25/100, and that reduces to 1/4.

Adding and Subtracting Decimals   Adding and subtracting decimals is easy as long as you line them up correctly. Consider this example: 1.287 + 24.32. First, be sure the two numbers have the same number of spaces after the decimal point. The first number has three places, but the second only has two. But, you can always add zeros to the end of a decimal number, 24.32 = 24.320. Note that the last 0 means there are 0/1000, which is correct and does not change the value. Finally, it is easiest to add these numbers if you line them up vertically, always being sure the decimal points are aligned:

$$\begin{aligned} 1.287 \\ + \ 24.320 \\ \hline = \ 25.607 \end{aligned}$$

Subtraction is done the same way.

PICTURE THIS

Assume that you have worked three extra jobs for some extra cash this week. From them, you earned £33.70, £45.28 and £21.02. How much extra money did you earn? _____

With this money you buy a pizza for £8.99, soft drinks for £1.49, petrol for £20.00 and a new CD for £19.95. How much money do you have left? _____

Congratulations, you have just added and subtracted decimals, as you do on a regular basis in daily life! You should see that you earned £100 and have £49.57 left.

When adding or subtracting decimals, always align the decimal point in the two numbers before doing the operation. ▪

Multiplying and Dividing Decimals   Multiplication and division of decimals is a bit trickier because you must keep track of how many decimal places you should have at the end. Let's try an easy one: 0.5×0.3. First, multiply the numbers as if they are whole numbers: $5 \times 3 = 15$. Now, add the number of decimal places you started with. Both numbers you multiplied originally had one decimal place, so that adds up to two. Realise that your answer of 15 is 15.0, so you know where the decimal begins. Now you have to move the decimal point. The numbers you started with had a total of two digits after the decimal point, so you must move the decimal point left by two places, giving you 0.15. Here is a way to double check that. If the original numbers were fractions, they would be 3/10 and 5/10. Recall how to multiply fractions – you multiply the numerators, then multiply the denominators:

$$3/10 \times 5/10 = 15/100 = 0.15$$

What if the numbers had been 0.03×0.5? Although you still get 15, now you need to move the decimal point three places to the left, but there are only two. You simply add zeros to the left until you have the correct number of decimal places, in this case giving you 0.015.

Division with decimals is just like ordinary division, except we keep going until we either finish or reach a predetermined stopping point. You usually do not go beyond the number of decimal points your original numbers contain, so if they had a total of two, you would probably stop at two and round off beyond that. Let's look at an example: $2.8 \div 7$. Put the decimal point in the answer line exactly above its position in the dividend (2.8), then simply do the division:

$$
\begin{array}{r}
0.4 \\
7)\overline{2.8} \\
-\ 2\,8 \\
\hline
0
\end{array}
$$

How do you divide when both numbers are decimals? Let's divide 2.1 by 0.7. All you have to do is move the decimal of the divisor until you have a whole number, then move the decimal of the dividend by the same number of spaces in the same direction. If you try 1.68 (dividend) $\div$ 0.3 (divisor), you move the decimal in 0.3 one spot to the right to get 3, then you must also move the decimal in 1.68 one spot to the right, getting 16.8, so the problem becomes $16.8 \div 3$. Do long division to get the result:

$$
\begin{array}{r}
5.6 \\
3)\overline{16.8} \\
-\ 15 \\
\hline
18 \\
-\ 18 \\
\hline
0
\end{array}
$$

You can always multiply back to double check your result:

$$
0.3 \times 5.6 = 1.68
$$

Moving the decimal point may seem confusing, but there are some easy shortcuts to remember.

- Moving the decimal point to the *right* one space is the same as multiplying by 10; 2 spaces multiplies by 100; and so on, so the numbers get *bigger*.

- Moving the decimal point to the *left* means you are dividing by 10 for each space moved, and the number always gets *smaller*.

LOOK OUT

It is important that when you read a prescription chart that you are certain of the position of the decimal point. Changing the decimal point by one place could mean that the patient has ten times the intended dose or has had a tenth of the dose prescribed. Both of these are drug errors, which can have serious consequences for the patient.

Moving the decimal point to the right is the same as multiplying by 10 for each space moved. Moving it to the left is the same as dividing by 10 for each space moved. ▧

✔ **QUICK CHECK**

Solve these problems:

1. $1.27 + 3.6 =$ _____

2. $14.87 - 3.2 =$ _____

3. $2.4 \times 1.2 =$ _____

4. $8.4 \div 0.2 =$ _____

Answers: 1. $1.27 + 3.60 = 4.87$. 2. $14.87 - 3.20 = 11.67$. 3. $2.4 \times 1.2 = 2.88$. 4. Move the decimal in the divisor and dividend both one space to the right, so it becomes $84 \div 2 = 42$.

Percentages

As you learn anatomy and physiology, you will encounter many physiological values that are stated as percentages. Per cent is another way of writing a fraction but is specific since the whole unit is divided into 100 parts. Each part is 1 per cent (%) or $\frac{1}{100}$ of the whole. For that reason, when working with percentages, always be sure they add up to 100 and no more than that.

Percentages are easy to work with. They are essentially fractions expressed as hundredths. For example, 25 per cent is the same as 25/100,

TABLE 2.1 The relationship between percents, decimals, and fractions.

Percentage	Decimal	Fraction in hundredths	Reduced fraction
10	0.10	10/100	1/10
25	0.25	25/100	1/4
40	0.40	40/100	2/5
50	0.50	50/100	1/2
75	0.75	75/100	3/4
100	1.00	100/100	1

which can be further reduced to 1/4, and because fractions can be expressed as decimals, so can percentages. You simply put the decimal two places to the left of the per cent (dividing by 100). So, 25 per cent becomes 0.25 and 7 per cent becomes 0.07. Percentages, decimals and fractions are all interchangeable, but when doing maths operations, percentages should be converted into either decimals or fractions. You cannot do maths operations with mixed expressions – they must all be whole numbers, or fractions, or decimals. **Table 2.1** explains the relationship between these expressions.

LOOK OUT

Oxygen is carried from the lungs to the tissues by the red blood cells. Normally, the amount of oxygen or saturation is close to 100 per cent. When a patient has breathing difficulties the amount of oxygen is reduced and so the percentage saturation is monitored. If it falls to only 92 per cent saturation, the patient is likely to need nursing and/or medical intervention.

APPLYING THE THEORY

Fluids which patients receive via an intravenous infusion (drip) are in concentrations that are expressed as a percentage, e.g. 0.9% saline or 5% glucose i.e., 0.9 g in 100 ml or 5 g in 100 ml. What concentration of sodium or glucose does each fluid contain? When you are out in practice, observe what other strength solutions are given.

TIME TO TRY

Can you supply the missing information in this table?

Percentage	Decimal	Fraction
36	_____	_____
_____	0.42	_____
_____	_____	80/100 = 4/5

Answers: 36% = 0.36 = 36/100 = 9/25; 42% = 0.42 = 42/100 = 21/50; 80% = 0.8 = 80/100 = 4/5.

Can You Feel the Power? **Understanding Exponents**

We briefly discussed exponents earlier, when we stated that a number written with an exponent is basically a multiplication problem: $2^3 = 2 \times 2 \times 2 = 8$. In science, very large and very small numbers are often written in a special format that uses exponents based on powers of 10. This format is called **scientific notation**. Let's consider the number 200 to see how this format is used. To write a number in scientific notation, first place the decimal point immediately after the first digit, and then drop the zeros. This number is called the **coefficient**. In this example, the coefficient is 2. Next, count how many spaces you moved the decimal – two places to the left. Each of those spaces represents a power of 10, so two places means $10 \times 10 = 100$. In scientific notation, we would write 200 as 2×10^2. As you can see, this is $2 \times 10 \times 10 = 200$. Remember, the first number in scientific notation must be greater than 1 but less than 10. If a number is less than 1, the exponent is a negative power of 10. For example, 0.0004 would be 4×10^{-4} because the decimal was moved four spaces to the right. You will rarely need to do maths operations with scientific notation, so we will skip those. You should, however, understand scientific notation so that you can understand some of the measurements you will read about – such as cell sizes, which are often measured in micrometres (10^{-6} metre). **Table 2.2** lists some common exponents.

TABLE 2.2 The values of some common exponents used in scientific notation.

Exponent	Value	Term
10^9	1 000 000 000	billions
10^6	1 000 000	millions
10^3	1 000	thousands
10^2	100	hundreds
10^1	10	tens
10^0	1	ones
10^{-1}	1/10	tenths
10^{-2}	1/100	hundredths
10^{-3}	1/1 000	thousandths
10^{-6}	1/1 000 000	millionths
10^{-9}	1/1 000 000 000	billionths

TIME TO TRY

Express the two numbers below in scientific notation.

24 000 000 = _____

0.003 = _____

If you did this correctly, you got 2.4×10^7 and 3×10^{-3}.

APPLYING THE THEORY

Having a number with a large number of 0s can be confusing to read and one could be added or missed when it is written, therefore we use scientific notation. You will see these numbers used on blood reports, e.g. Red blood cells = 5.5×10^{12}/l which would otherwise have to be written as 5 500 000 000 000 cells per litre! (Count the number of places, from the last 0, that the decimal point has been moved.)

My Cell's Bigger than Your Cell! **Ratios and Proportions**

- Sodium and potassium move across cell membranes in a 3:2 relationship.

- In the United States, the ratio of males to females at birth is about 105:100.

- The ratio of males to females declines steadily until, after age 85, it is only 40.7:100.

Welcome to the comparatively interesting realm of ratios. A **ratio** expresses a relationship between two or more numbers – it is a way to compare them. Ratios can be expressed using a colon between the numbers (as above), as a fraction, or by using the word 'to'. For example, carbohydrates contain hydrogen and oxygen in a 2 to 1 ratio, meaning there are twice as many hydrogen atoms in carbohydrates as there are oxygen atoms.

Ratios are used for comparison, and they can also be expressed as fractions. For example, a ratio of 1:2 means the same as 1/2. Look at that carefully, though. Let's say the ratio of men to women in your anatomy class is 1:2. We're not saying that half of the class are men, we are saying there are half as many men as women.

If ratios can be expressed as fractions, they can also be expressed as decimals and percentages. Because they can be written as fractions, they can also be reduced like fractions. For example, a ratio of 4:6 is the same as 4/6, which is the same as 2/3. When working with ratios, it is critical to write them in the correct order. If an anatomy class has 10 males and 20 females, the ratio of males to females is 10:20. If we write it as 20:10, it means there are twice as many males as females, which is not true.

TIME TO TRY

Empty your pocket or purse of change. Separate the coins by denomination. Count all of the coins in each category. Now express those numbers in a ratio: _____

Why is it important to indicate the order in which you are listing the coins? _____

If you did this correctly, you should have indicated the order of the coins, because without that reference, we have no idea what number corresponds with which coin. Perhaps you had 5 pennies, 4 20p pieces 3 50p coins, and 2 £1.00 coins. If you wrote your ratio in that order, it would be 5:4:3:2.

We also use ratios to discuss quantities in a certain amount. For example, there are about 280 million haemoglobin molecules in each red blood cell. That can be expressed as 280 million/cell, which looks more like a fraction, but it is really saying the ratio is 280 million to 1. Rates are also a special type of ratio. A red blood cell travels about 700 kilometres in its 120-day lifespan, giving a ratio of 700 kilometres/120 days. Drug doses are also often given in this rate format; for example, 5 milligrams per kilogram of body weight.

Proportions are statements of equal ratios. A simple example would be to say $1/2 = 4/8$. In science, we often use proportions to solve problems. To see how, examine a generic version:

$$\frac{a}{b} = \frac{c}{d}$$

Because these two ratios are equal, their cross-products are also equal, due to some basic laws of maths. This means that the product of multiplying the first numerator (a) by the second denominator (d) equals the product of multiplying the second numerator (c) by the first denominator (b):

$$a \times d = c \times b$$

Let's say we want to know how many times the heart beats in an hour. Assume that the heart beats on average 80 beats per minute. We know there are 60 minutes per hour. So, we can set up the proportion, filling in the information we know and using 'x' to represent the value we are trying to determine:

$$\frac{80 \text{ beats}}{1 \text{ minute}} = \frac{x \text{ beats}}{60 \text{ minutes}}$$

We know we can **cross-multiply** (**Figure 2.4**). When we do that, we get $4800 = x$, so there are 4800 beats per 60 minutes, or per hour. In fact, there is a simpler way to write this problem, which is a shorter version of cross-multiplying. We know there are 80 beats per minute,

1. Set up the proportion with what is known, using 'x' to represent the information you are seeking.

$$\frac{80 \text{ beats}}{1 \text{ minute}} = \frac{x \text{ beats}}{60 \text{ minutes}}$$

2. Next, cross multiply to solve for 'x'.

$$\frac{80 \text{ beats}}{1 \text{ minute}} = \frac{x \text{ beats}}{60 \text{ minutes}}$$

80 × 60 = 4800, and 1x = x

So x = 4800 beats per 60 minutes, or 1 hour

FIGURE 2.4 Using proportions to determine how many times the heart beats in an hour.

and 60 minutes per hour, so we can calculate the beats per hour as follows:

$$\frac{80 \text{ beats}}{\text{minute}} \times \frac{60 \text{ minutes}}{\text{hour}} = \frac{4800 \text{ beats } \cancel{\text{minutes}}}{\cancel{\text{minute}} \text{ hour}} = \frac{4800 \text{ beats}}{\text{hour}}$$

Notice that the units are shown, and in the next to last step, minutes appear on both the top and bottom. That means they cancel each other out, so we are left with beats per hour, which is the correct unit. Using the units can be an easy way to ensure that you have set up the problem correctly. This is another example of taking the time to think through the problem before you start – the units should make sense when you are done.

✔ **QUICK CHECK**

An average person takes about 12 breaths per minute. How many breaths do they take in an hour? _____

Answer: Set up the proportion: 12 breaths/minute = x breaths/60 minutes. Cross-multiply: 12 × 60 = 720 = x, so x = 720 breaths per hour.

Who Ever Heard of a Centimetre Worm? **The Metric System**

The **metric system**, or **System Internationale (SI)**, is universally used in science and by almost every country in the world except the United States. You have undoubtedly had brushes with learning the metric system, and you may have found it difficult.

WHY SHOULD I CARE?

Science uses metric measurement almost exclusively, so you will need a basic understanding of metric units for your course work and for your future career. In addition, almost everyone on our planet – except the United States – uses the metric system.

The metric system is amazingly simple because it is all based on the number 10, so obviously decimals are easy. For our purposes, we'll learn four main units used in science: those that deal with length or distance, mass, volume and temperature. Each of these has a standard or base unit:

- The basic unit of length (or distance) is the **metre (m)**.

- The basic unit of mass is, technically, the **kilogram (kg)**, but many sources use the **gram (g)** as the base unit instead.

- The basic unit of volume is the **litre (l)**. (Sometimes litre is abbreviated as L to avoid confusion with the numeral 1.)

- The basic unit of temperature is the **degree Celsius (°C)**.

These are the base units, but more convenient units are derived from these. For example, a metre is just a bit longer than 3 feet (39.34 inches), so it is not a convenient unit for measuring the size of, say, your finger or a cell. Smaller units of the metre, based on the powers of 10, are used instead. These units are named by adding the appropriate prefix to the term *metre* (Table 2.3). **Centi-** means 1/100, and there are 2.54 centimetres (cm) in an inch, so centimetres work well for measuring fingers. Cells are microscopic, so they are best

TABLE 2.3   Metric system prefixes.

Prefix	Symbol	Decimal equivalent (multiple)	Exponential equivalent (Scientific notation)
Pico-	p	0.000000000001	10^{-12}
Nano-	n	0.000000001	10^{-9}
Micro-	μ	0.000001	10^{-6}
Milli-	m	0.001	10^{-3}
Centi-	c	0.01	10^{-2}
Deci-	d	0.1	10^{-1}
no prefix		1.0	10^{0}
Deka-	D	10.	10^{1}
Hecto-	H	100.	10^{2}
Kilo-	k	1000.	10^{3}
Mega-	M	1 000 000.	10^{6}
Giga-	G	1000 000 000.	10^{9}

measured in even smaller units, such as micrometres – one micrometre = 1 millionth of a metre. Driving between cities, you can best measure the long distances in kilometres, each of which equals 1000 metres. Again, all metric units are based on 10. Think about that – you first learn to count from 1 to 10, then you can count by tens to 100, then by hundreds to 1000, and so on. It is an easy system.

Table 2.3 provides many of the prefixes and their base-10 equivalent. In anatomy and physiology, you will use some units more often than others. For length or distance, which is a straight linear measurement, you will mostly work in metres, centimetres, millimetres (1/1000 m), and micrometres. For mass, which is the actual physical amount of something, you will most often refer to kilograms (1000 grams), grams and milligrams (1/1000 g). For volume, which refers to the amount of space something occupies, the most common units will be litres and millilitres.

The Celsius temperature scale does not use prefixes. Instead, it has a single unit: degrees Celsius. The scale is divided into 100 degrees, with 0°C being the freezing point of pure water and 100°C its boiling point.

PICTURE THIS

If you drink carbonated soft drinks, you are probably familiar with their standard large plastic bottles. What is their volume in metric units? _____ litres

Answer: 2 litres

Know the paperclip! A standard small paperclip has a mass of about 1 gram (it's very light). A standard large paperclip has a side-to-side width of about 1 cm, and the wire from which it is made has a diameter of about 1 mm.

You will become more familiar with the metric system as you use it. In class almost all measuring and discussion will use metric units.

Let's try some conversions. First, let's do the easy stuff: converting between metric units. Remember that the difference between the units will always be some multiple of 10. Let's convert 13 metres into centimetres.

$$13 \text{ m} = \underline{\hspace{3cm}} \text{ cm}$$

A centimetre is 1/100 of a metre, so there are 100 centimetres per metre. Thus:

$$13 \text{ m} \times 100 \text{ cm/m} = 1300 \text{ cm}$$

Now we will convert 27 millimetres into centimetres, but let's try another method. All we really have to do to convert between different metric units is move the decimal point, but by how many spaces and in which direction? How many spaces you move is determined by the difference in the power of 10. We know that millimetres are thousandths of a metre, and centimetres are hundredths of a metre.

$$\text{millimetres} = 10^{-3} \qquad \text{centimetres} = 10^{-2}$$

So, if we look at the exponents, they are different by one. We will move the decimal point in our number (27) by one place. But in which direction? When converting from smaller to larger units, the decimal point moves left. When converting from larger to smaller units, the decimal point moves right. Back to our example: converting 27 mm to cm gives us 2.7 cm.

If you have difficulty with thinking about which way to move the decimal point, just think about money. One pound is equal to one hundred pennies. A pound is written as £1 or £1.00. We do not usually put in a decimal point if we have whole numbers, but it is useful to remember that in theory it is still there. If you have 200 pennies saved in your piggy bank, it is unlikely that you would want to carry them in that form in your pocket, unless you are into weight training! The sensible thing would be to exchange them for two one-pound coins. What you are doing is changing the smaller units into bigger units, so you will have fewer of the bigger units. The number you end up with is smaller but has the same value. It is just the same with the units of length, mass and volume in the metric system, so when you are converting, just ask yourself 'Will I expect more or fewer units?' You should then know which way to move the decimal point.

In the example above you had 200 pence. You know that there are 100 pence in a £1.00, so you divide the number of pence by 100 – move the decimal point two places to the left to give you £2.00.

When converting within metric units:

1. Put the units in scientific notation and subtract the smaller exponent from the larger one. The difference is how many spaces the decimal point will move in your coefficient.

2. If you are converting from small units to larger ones, the number gets smaller, so the decimal point moves to the left. If you are converting from larger units to smaller ones, the number gets bigger, so the decimal point moves to the right. ■

LOOK OUT

A patient's fluid intake and output may be measured in millilitres (ml) throughout the day but the total will be converted to litres (l) at the end of 24 hours.

TIME TO TRY

Now that you see the simple secret to this process, complete the following conversions:

5 kg = _____ g   8 ml = _____ l   6 cm = _____ m

2 g = _____ mg   25 mg = _____ g

If you did these conversions correctly, you should see that 5 kg = 5000 g;   8 ml = 0.008 l;   6 cm = 0.06 m;   2 g = 2000 mg;   25 mg = 0.25 g. See, the metric system is easy!

How Do You Measure Up? **Basic Measurement**

Now that you understand the basic units of measurement, you need to know how to measure. A common error in scientific experimentation is called human error, which includes maths mistakes (which you won't make now!) and something as simple as not measuring correctly. When you're baking brownies, adding extra sugar and chocolate may be a good thing, but that won't work in science. Measurements must be made precisely and with appropriate equipment.

Measuring Length

Length is usually measured with a metre stick or a ruler. Grab one – surely you can locate one somewhere. I'll wait . . .

Examine the scale: there are probably two – inches and metric. On a metre rule, the scales are often on opposite sides of the rule. Look at the metric scale. If it is a metre rule or metric tape measure, the small numbered units are usually centimetres. Confirm that there are 100 of these in a metre. Compare the size of a centimetre to an inch by placing your fingers on each side of a centimetre, then maintaining that space as you move to the inch scale. Note that an inch is a bit over 2.5 times bigger – 2.54 to be exact (**Figure 2.5a**).

Now examine the space between 0 and 1 cm. Count the spaces. How many are there? _____ You should have counted 10 spaces. These tiny units are millimetres, each equal to 1/1000 metre. That makes sense – each millimetre is 1/10 of a centimetre, which is 1/100 of a metre, and 1/10 of 1/100 is, indeed, 1/1000.

When measuring with a ruler, be sure to align the zero line exactly at the edge of your object and measure exactly to the far edge. Do not round off. If you are measuring the diameter of a circle, be sure your ruler is positioned across the widest part of the circle (**Figure 2.5b**).

When measuring length, be sure you use the appropriate scale. ▪

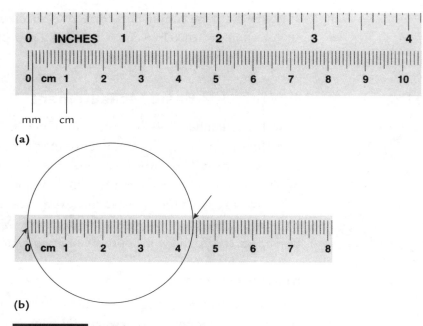

FIGURE 2.5   Using a metric ruler. **(a)** This shows the comparison of the metric and inch scales. Millimetres and centimetres are labelled. **(b)** Always start measuring at the very edge of the object, positioning the ruler carefully so that the scale runs precisely through the dimension being measured.

Measuring Mass

Mass is the actual amount of something, and it is closely related to weight, but weight takes into account the force of gravity acting on the mass. Mass will be constant, but weight will vary with gravity – just ask the astronauts, who have no weight in outer space where there is no gravity. Here on Earth, where the gravitational pull is rather steady, the terms mass and weight are often used interchangeably, but you should realise that they are different.

Mass can be measured with a digital scale that measures in grams, but often in labs we use the triple beam balance (**Figure 2.6a**). Before using the balance, be sure all the attached standard masses are pushed as far to the left as possible. Next, you must 'zero' the balance before adding anything to the pan. Notice how the two lines on the far right line up. One line is on the arm and moves with it; the other is on the end piece of the scale. If they are not perfectly aligned, slowly turn the

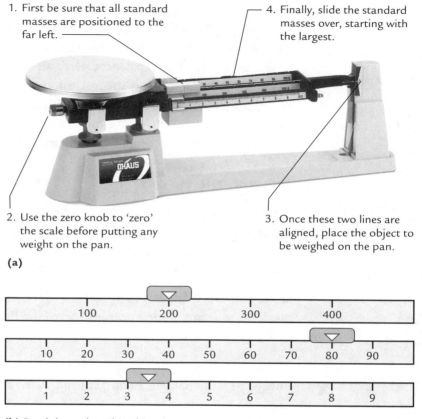

1. First be sure that all standard masses are positioned to the far left.

4. Finally, slide the standard masses over, starting with the largest.

2. Use the zero knob to 'zero' the scale before putting any weight on the pan.

3. Once these two lines are aligned, place the object to be weighed on the pan.

(a)

(b) Read the scales. This object has a mass of 283.5 g.

FIGURE 2.6   Using the triple beam balance.

zero knob located on the left of the scale, usually under the pan. Rotate the knob in either direction until the two lines are aligned. The arm will move up and down a bit – wait until it has stopped and the lines are aligned. The scale is now zeroed. NOW you are ready to measure the mass of your object.

There are three beams on the arm of the balance, each suspending a different standard mass. The largest mass is 100 g. Each spot that you move that mass to the right equals 100 g. Another beam has a 10-g mass, and the front beam has a 1-g mass. Starting with the 100-g mass, slide it across until it causes the arm to swing too far to the right. That was too heavy, so back it up to the left by one spot. Note the number – that is how many hundred grams are in your object. Next, slide the

10-g mass until it is too much, back up one notch, then note how many 10s of grams there are. Finally, slide the 1-g mass carefully until the two lines on the right side again align, as they did at the beginning. At that point, the scale is balanced. Read all of the whole numbers on the scale; each line beyond the last whole number is 1/10 g. If the 100-g mass is at 200, the 10-g mass is at 80, and the 1-g mass is halfway between the 3 and 4, what is the mass of the object? _____

(It would be 283.5 g, as shown in **Figure 2.6b**.)

APPLYING THE THEORY

Baby weighing scales use the same principles except that the graduations will be in kilograms and grams, and the weighing pan curved to keep the infant safe. Patience is needed to get an accurate reading – wriggling disturbs the balance!

Some adult weighing scales are of a similar construction, using a platform to stand on or a seat instead of the flat pan.

Always zero the balance before starting, with all standard masses at the far left. ▪

Measuring Volume

Volume refers to the amount of space a substance occupies. In anatomy labs, you will most often measure liquid volumes by using beakers or graduated cylinders. Always use the smallest container in which the substance will fit – the smaller it is, the more accurate your reading will be. Always read the scale, usually in millilitres, at eye level for accuracy. You should also know how to read the meniscus (**Figure 2.7**) – this is especially critical in a graduated cylinder. Liquid in a container tends to climb slightly up the sides. This is due to molecules of the liquid 'clinging' to each other and the sides of the container. This is called surface tension. The centre is lower than the edges, where the liquid contacts the container. This dip is called the **meniscus**. When reading the scale, always read it at the low point of the meniscus for the best accuracy.

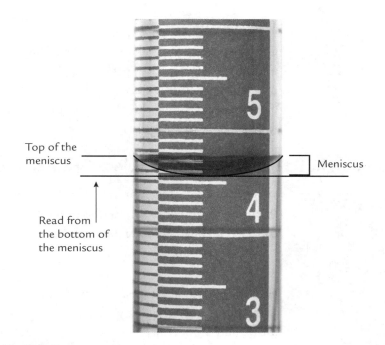

Top of the meniscus

Meniscus

Read from the bottom of the meniscus

FIGURE 2.7 Read measurements of liquid volumes from the bottom of the meniscus. By doing so, you see that this graduated cylinder contains 4.55 ml of liquid, not the 4.75 indicated by the top of the meniscus.

Source: © Richard Megna, FUNDAMENTAL PHOTOGRAPHS, NYC.

LOOK OUT

If you draw up drugs into syringes or pour liquid medicines into measuring cups it is vital that the measurements are accurate. You must read the meniscus at eye level to ensure accuracy. Pouring medicines into spoons is also influenced by surface tension – you can get a heaped spoonful if you are not careful. A 5 ml dose is level with the rim of the spoon.

TIME TO TRY

Find the narrowest clear container you can and fill it halfway with water. Look at it at eye level. Use a ruler on the outside of the container to measure the highest point of the water. _____
Now measure the lowest point. _____ The dip that you see is the meniscus. Whenever you measure a liquid volume, always measure at the lowest point of the meniscus.

TIME TO TRY

Get a teaspoon and fill it with water. Then add a drop at a time and watch how you can get a heaped spoon of water. How many drops can you add before it spills over? Try the same procedure using washing-up liquid. Does the spoon hold more or less liquid?

You Ought to be in Pictures: **Tables, Graphs and Charts**

We explored how to get numbers by measuring and how to work with them. Now we will see how these numbers and other information, collectively called data, can be depicted.

Tables

We have already used tables in this book, and you should be familiar with them, so let's quickly review the basics. Tables come in many forms and are a convenient way to present information so that it is easy to read and compare. We will use **Table 2.4** as a reference.

When viewing a table, start with the table title, in this case 'Fluctuations in body temperature throughout the day.' The title usually tells you what the table contains. Note that tables are carefully arranged in columns and rows. All information in a single column is related, and all information in a single row is related. Look at the top of each column – these are column heads that tell you what information each column contains. Look at the beginning of each row. These are row heads, or labels that tell you what each row contains. In our example, the column heads reveal that the first column identifies each test subject, and the other columns contain the temperatures for all test subjects at specific times of the day. The row heads tell us that all temperatures in

TABLE 2.4   Fluctuations in body temperature throughout the day.

Subject	T °C at 4 A.M.	T °C at 8 A.M.	T °C at noon	T °C at 4 P.M.	T °C at 8 P.M.	T °C at midnight
A	36.0°C	36.3°C	36.4°C	36.8°C	37.0°C	36.8°C
B	38.0°C	38.4°C	38.6°C	38.8°C	39.0°C	38.4°C
C	37.6°C	38.0°C	38.4°C	38.5°C	38.9°C	38.6°C
D	37.2°C	37.3°C	37.5°C	37.6°C	38.0°C	37.6°C

a single row belong to a single test subject, and who it is. So, by using the column and row heads, it is easy to find out, for example, what temperature Subject C had at noon.

TIME TO TRY

On a separate piece of paper, using sentences and paragraphs, write out all the information that is included in Table 2.4. Which version is easier to read? In which version can you more easily determine Subject D's temperature at 8 P.M.? _____

Graphs

Graphs present a more pictorial view of data. Numerical data that can be organised in a table (**Figure 2.8a**) can usually also be presented in a graph. The main advantage is that the graph allows you to spot trends and relationships almost instantly. There are various types of graphs. Let's look at three of them: line graphs, bar graphs and pie charts.

Line Graphs   Look at **Figure 2.8b**. This is a line graph, and they usually are laid out in a grid. The horizontal axis at the bottom is the

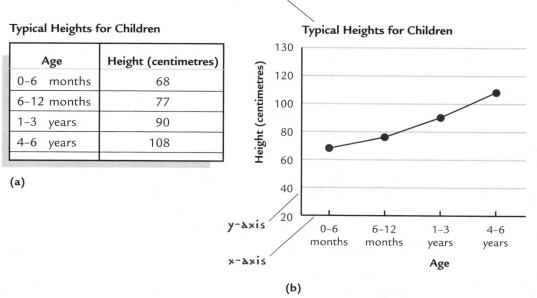

Typical Heights for Children

Age	Height (centimetres)
0–6 months	68
6–12 months	77
1–3 years	90
4–6 years	108

(a)

(b)

FIGURE 2.8   Typical heights for children up to 6 years of age. **(a)** Table format. **(b)** Basic line graph.

x-axis. It often, but not always, marks the progression of time. The vertical axis on the left is the *y*-axis and it typically reflects some increasing value. Where these lines meet on the lower left of the graph marks the 0 position, so units go up as you move away from that point. Each axis should be labelled and should include units.

In our example, the *x*-axis tells us who the data are about – the age group of children for whom the listed height is typical. The *y*-axis gives the typical height for each age group, in centimetres. Each height (data point) is placed above the age it represents. The data points may be left unconnected, they may be directly connected (as they are here), or a line of 'best fit' may be drawn that passes between the points so they are evenly distributed on each side of it. In our example, because this depicts growth with time, the line allows us to see at once the heights for each age group and to quickly comprehend the trend of a gradual increase in height with age.

When drawing a line graph, don't forget to label the axes and to include units. The most common mistake made when graphing data is to not use an appropriate scale. Be sure to size the units to maximise the space the graph fills. You don't want the graph to be cramped into one corner, making it hard to read. Spread it out both vertically and horizontally.

LOOK OUT

When you are reading graphs and charts in publications, make sure you look carefully at the scales used. An unequal distribution on each scale can distort the perception of the information. (You can fool some of the people, some of the time!)

Bar Graphs   **Figure 2.9** is a bar graph showing the amount of water in the human body. Looking at the axes, you see that the *x*-axis has three separate categories: total water, intracellular fluid (the water located inside body cells) and extracellular fluid (the water not contained in cells). Each of these categories has two bars – one for males and one for females. The *y*-axis tells what percentage of the total body weight the water represents. This graph is drawn so that the male and

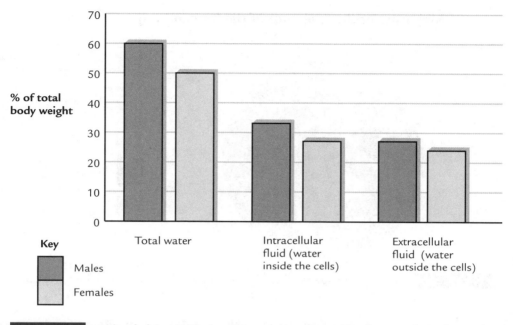

FIGURE 2.9   A simple bar graph showing a comparison of body water in males and females.

female data are directly compared by being positioned side by side, yet easily distinguished by use of different shading. The shading is explained on the lower left in a feature called the **key**. This bar graph has a 3-D effect that does not change its meaning at all – it just makes it slightly more interesting. Bar graphs may be drawn vertically or horizontally. Because each bar is so distinct, these graphs are good for comparing specific bits of data, whereas line graphs may be better at showing an overall trend.

Pie Charts   **Figure 2.10** is a pie chart, a type of graph that shows parts of a whole. This one shows the main molecules that make up the human body and tells the percentage of the body made of each type of molecule. When looking at this, you immediately see that all the parts add up to the whole 'pie', which is 100 per cent, so these charts are very effective when showing percentages. We all have a visual concept of a whole pie and a slice of pie, so even before looking at the numbers, we instantly see the differences in distribution. You know right away from this pie chart that one type of molecule makes up well over half of the body.

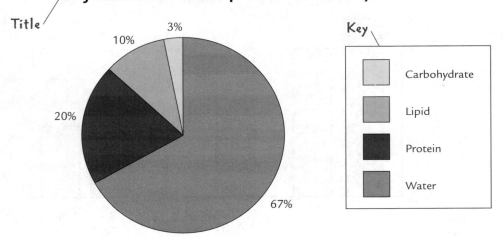

Major molecular make-up of the human body

Title

3%

10%

20%

67%

Key

Carbohydrate

Lipid

Protein

Water

FIGURE 2.10   A pie chart showing the major molecular make-up of the human body.

Pie charts have no axes to label and the space within them is limited, so a key is often used. Labels for each 'slice' may be written within the pie or placed to the outside, as in our example. Using different shading or colours for the slices makes these graphs even more readable. Because there are no axes to provide information, the graph title and key are very important.

When reading a graph, always read the title first, then the axes, key and labels. Finally, just let your eyes take in the relationships depicted. ■

✔ **QUICK CHECK**

1. In Figure 2.8, which age group is shortest? _____ What is the typical height for children aged 1 to 3 years? _____

2. In Figure 2.9, what percentage of the total body weight is water in an average female? _____

3. In Figure 2.10, which molecule makes up most of the human body? _____

Answers: 1. The shortest group is aged 0–6 months, and children aged 1–3 are typically 90 cm tall. 2. 50 per cent. 3. Water.

FINAL STRETCH!

Now that you have finished reading this chapter, it is time to stretch your brain a bit and check how much you have learned.

RUNNING WORDS

At the end of each chapter, be sure you have learned the language. Here are the terms introduced in this chapter with which you should be familiar. Write them in a notebook and define them in your own words, then go back through the chapter to check your meaning, correcting as needed. Also try to list examples when appropriate.

Data	Ratio
Product	Proportion
Exponent	Cross-multiply
Dividend	Metric system (SI)
Divisor	Metre (m)
Quotient	Kilogram (km)
Normal	Gram (g)
Average	Litre (l)
Mean	Degree Celsius (°C)
Numerator	Centi-
Denominator	Milli-
Equivalent fraction	Kilo-
Reduce	Mass
Greatest common factor (GCF)	Volume
Common denominator	Meniscus
Least common multiple (LCM)	x-axis
Repeating decimal	y-axis
Scientific notation	Key
Coefficient	

WHAT DID YOU LEARN?

Try these exercises from memory first, then go back and check your answers, looking up any items that you want to review. Answers to these questions are at the end of the book.

PART A: SOLVE THESE PROBLEMS

1. $(4 \div 2) + 6 - 5 \times 2^3 =$ _____

2. $2 \times 10^4 =$ _____

3. 27/36 reduced is _____

4. $3/8 \times 2/3 =$ _____

5. $5/6 - 7/12 =$ _____

6. $0.5 \times 0.4 =$ _____

7. If the respiratory rate is 12 breaths per minute, how many breaths are taken in 1 hour? _____

8. 4 metres = _____ centimetres

9. 7.5 g = _____ mg

10. If a man weighs 75 kg and 60 per cent of his body weight is water, how many kg of water does he have? _____

PART B: ANSWER THESE QUESTIONS

1. What is the mean of 8, 9, 12, 18 and 23? _____

2. What is the numerator in 4/5? _____

3. Express 3/10 as a decimal _____ and as a percentage _____

4. From Figure 2.10, how much of the human body is made of protein? _____

5. What would you be measuring if you are looking at a meniscus? _____

6. In the metric system, list the base unit for each of the following:

mass: _____

length: _____

volume: _____

7. What is meant by 'normal' blood pressure? _____

8. At what Celsius temperature does water boil? _____

9. The amount of space something occupies is called _____.

10. How many milligrams are there in 1 gram? _____

WEB RESOURCES

Here are some additional online resources for you.

- *Maths centre*

http://www.mathcentre.ac.uk/students.php/

A number of resources produced by a group of people from various higher education institutions to assist in the study of important areas of pre-university mathematics. A pick and mix menu for you to revise areas that are rusty and skip those where you are confident.

- *BBC websites*

http://www.bbc.co.uk/skillswise/numbers/

There are a number of sites to assist with revision. Try the skillswise website, if you need to brush up on your addition and subtraction. There are also exercises and explanations of fractions, decimals and percentages. The section on handling data is useful for ensuring you know how to read graphs and tables, especially for understanding the research papers you will be expected to read and critique during your course.

http://www.bbc.co.uk/schools/gcsebitesize/maths/

Again, you can look at the areas you need to revise and you can work from the foundation to the intermediate level, which is roughly the minimum standard you will need to reach.

http://www.bbc.co.uk/asguru/maths/

If you are reasonably happy with your knowledge and would like to push yourself a bit further, look at the site for A-level revision.

■ *Fun and relaxation with a purpose*

http://www.easymaths.com

This site will help to convince you that maths can be fun as well as honing some basic skills.

■ *Headstart in biology*

http://learninglab.co.uk/headstart/

or

http://www.headstartinbiology.com

This site has material related to several chapters in this book. The first gallery on measurement is particularly relevant for this chapter and will help you to see why getting to grips with arithmetic is so important.

The site is an introduction to biology for all healthcare students. It was originally designed for students who had just heard that they had been accepted onto a nursing course, and who wanted to use the few weeks before the course began to build a foundation of useful biological concepts. Consequently, the language and references in this resource highlight nursing situations rather than examples from other professions. However, the material and concepts explored are relevant to a range of healthcare professionals and so a broad range of students will be able to use this resource effectively.

Look at this free website – you will need to register.

3 Terminology

You Say Humerus, I Say Funny Bone

When you have completed this chapter, you should be able to:

■ Break down medical terms to understand their meanings.

■ Build medical terms from basic roots, prefixes and suffixes.

■ Describe the anatomical position and understand its relevance.

■ Correctly use basic anatomical terms to describe the relative positions of body parts, planes and sections and general body regions.

YOUR STARTING POINT

Answer the following questions to assess your knowledge of anatomical terminology.

1. Most anatomical terms arise from which languages? _____

_____

2. What is the difference between *anatomy* and *physiology*? _____

_____

3. What is meant by the term *cardiomyopathy*? _____

_____

4. What is meant by the *anatomical position*? _____

_____

5. What is meant by the term *medial*? _____

_____

6. What plane passes through the body from front to back? _____

_____

How Do You Say, In Your **Language** . . . ?

Have you ever read a computer manual only to find that you are still unsure of how to configure your personal firewall, or for that matter why you should? Has your mind gone numb as an auto mechanic explained all those expensive malfunctioning parts that were replaced to get rid of that mystery noise in your car? Have you tried to read the small print in an advertisement for a new prescription drug? At such times, it may seem as if other people speak a secret language. Indeed, most professions have their own sublanguage, as do most academic disciplines. Anatomy and physiology certainly are not exceptions.

Answers: 1. Latin and Greek. 2. Anatomy studies the body's structures; physiology studies their functions. 3. Disease of the muscle of the heart. 4. The body position used as a common reference point – standing erect, everything facing forward, including palms. 5. Toward the midline. 6. Sagittal.

Learning the terminology in these two disciplines may be a challenge, but I won't say it is difficult. *Difficult* leaves the option of ducking behind excuses. I have often heard my own students say, 'I could do this if the words weren't so hard, but this is just too difficult for me!' Instead, you should think of the words as a challenge, because a challenge sparks the competitor in us – it makes us try harder. Although the words may seem intimidating at first, you will quickly learn tools and tricks to help you understand even the longest terms. Are *you* up for the challenge?

PICTURE THIS

Upon arriving at your long-anticipated vacation destination, you notice how wonderfully new and exotic everything seems – the smells in the air, the people's faces and clothing, the food sold from vendors' carts, and the sounds. You are in paradise! Then you discover that both your luggage and your wallet are missing. You approach an authority and start explaining your situation only to be met with a blank stare. You suddenly realise part of the exotic sounds you have been hearing are a very different spoken language.

1. How will you convey your plight to this person? _____

   _____

2. How difficult would it be if you could speak the local language?

   _____

3. What could you have done before your trip to avoid this communication gap? _____

   _____

Obviously if you plan to spend time in a place where another language is spoken, communication may be a challenge unless you learn the local language in advance. Your anatomy and physiology class will be a place where another language is spoken. Much of the terminology will be new, and most of the words do, in fact, come from languages other than English. You will learn about body parts and their functions by listening to lectures filled with this new language, and you will be expected to discuss your course material and write exam answers using this new language. That is why it is so important that you learn the language of anatomy and physiology. As you study, the first thing you should do is master – not just read, but *master* – the new words. You have to know the language before you can understand and join in the conversation.

Learning the vocabulary is the first step in learning anatomy and physiology. ▪

I recommend that you keep a running vocabulary list. The easiest way to do this is to keep a separate notebook into which you write all new terms as they are introduced (**Figure 3.1**). If you write new terms in your notes during lectures, transfer them to your vocabulary list and be sure to check their spellings later. Once you know that a term is spelled correctly, write down the actual definition – exactly what does the word mean? You can find the actual textbook definition in your textbook or in a medical dictionary, but you should also try to explain the term in your own words.

Next, be sure you can use the word properly in a sentence. Try to add some examples that illustrate the term, if appropriate. For example, a tissue is a group of cells organised together that share a common function. Examples of tissue include, bone, blood, cartilage, fat and muscle. If you are a visual learner, try illustrating your new words if you can. If you are an auditory learner, try reading the words and their meanings out loud and consider tape recording your list. If you are a tactile learner, you might benefit from typing your list into a word-processing file, then putting them in alphabetical order. You might also write the terms and their meanings on flash cards. Make your own and have quizzes with your learning group. (A prize always increases motivation!)

You may be surprised at the beginning of the course at the number of new terms you encounter. You can think of the words as a kind of smoke screen – the underlying principles of anatomy and physiology are rather simple, but you may not see that through the smoke. More than one student has jokingly commented to me that 'It's all Greek to me'. They don't realise that their joke is not far from the truth – most anatomical terms have either Greek or Latin origins, although many

My Vocabulary List

Anatomy = The study of body structures.
Physiology = The study of body functions.
Biology = The study of life.
Cytology = The study of cells.
Histology = The study of tissues.

Nucleus = 1. The central region of an atom (in chemistry), where the protons and neutrons are located.
OR
2. The cell organelle that houses the DNA, found only in eukaryotic cells.

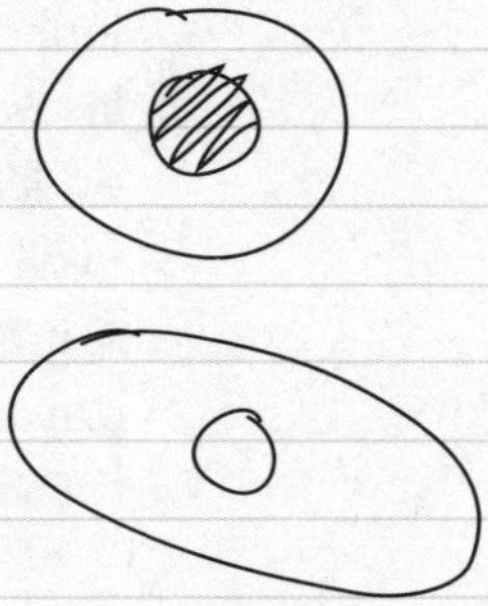

Ribosome = The cell organelle at which proteins are constructed (process of translation).
Mitosis = Division of a cell's nucleus.
Cytokinesis = Division of a cell's cytoplasm.

FIGURE 3.1 Keeping a separate running vocabulary list throughout your course of study can help you master the language.

terms are derived from other languages as well. So, instead of thinking that it is hard to learn the science because of all the big words, look at it this way: while learning the science, you will learn some new languages!

✔ QUICK CHECK

From which languages do most anatomical terms originate?

Answer: Greek and Latin.

The More the Merrier . . . **Types of Terms**

When studying anatomy and physiology, you will encounter many different types of terms, many of which you will already know. Ideally, all terms used in anatomy and physiology would have the same origin and follow the same rules, but in reality, of course, that is simply not true.

Descriptive Terms

Most terms you will encounter are descriptive, meaning the name is closely related to its meaning. For most students, these words often look the most intimidating – they can be quite long. Yet, once you know some simple rules, **descriptive terms** become quite easy to understand and deciphering their meanings actually becomes fun. Even large words like *electroencephalography* become manageable with some practice, and just think how empowered you will feel when you can actually converse with a doctor in his or her own language! We will spend most of our time in this chapter working on descriptive terms.

Eponyms

Another type of term is the **eponym**, which literally means 'putting a name upon'. Eponyms are terms that include someone's name and have been used traditionally to honour the person who first described a certain structure or condition. This practice led to terms such as the:

- Islets of Langerhans in your pancreas, first described by the German pathological anatomist Paul Langerhans;

- Sphincter of Oddi, the 'door' where pancreatic juice enters your small intestine, named after the Italian anatomist and surgeon Ruggero Oddi; and

- Eustachian tubes between your ear and your throat, named after the Italian anatomist Bartolomeo Eustachi.

Obviously, these terms are not so easily understood. Few people now associate an eponym with the discoverer, making the terms harder to master. Recent practice in anatomy has moved away from eponyms to a preference for descriptive terms, so those will be our focus. You will learn relevant eponyms as you move through your course, but fewer and fewer are in use today.

Of course, there are exceptions. Down's syndrome was previously referred to as *mongolism* because the characteristic eye appearance of someone with Down's syndrome was a bit like that of people of Asian descent. This term, however, is deemed offensive, so the eponym is now used more often. Another descriptive term for this genetic disorder is trisomy 21, which tells us that three copies of chromosome 21 cause this condition. Although few people know what trisomy 21 means, most people recognise and understand the eponym.

Most of the terms for your course are part of a broader area known as *medical terminology.* Eponyms are still used relatively commonly in medical terminology and in health-related fields, especially for naming diseases or abnormal conditions. You will occasionally encounter them, and you really just need to memorise them as they come up.

TIME TO TRY

Parkinson's disease is also called *paralysis agitans.*

1. Which is the descriptive term? _____

2. Which is the eponym? _____

3. With which term are you more familiar? _____

4. The first vertebra of your spine, located just under your skull, is also called the *atlas,* which is an eponym. Either recall who Atlas was in Greek mythology, or look up his name. How is this eponym also a descriptive term? _____

_____

Although paralysis agitans is the descriptive term, you are probably more familiar with the eponym – Parkinson's disease. You also probably realise that the atlas (bone) holds up your head, similar to the way Atlas from mythology held up the world.

Abbreviations and Acronyms

Like many disciplines, the worlds of anatomy and physiology are full of shortcuts. After all, some of those descriptive terms grow quite large! You will often encounter modifications of full terms in the form of abbreviations or acronyms. You are already familiar with abbreviations. An **abbreviation** is a shortened form of a word or phrase. For example, the part of your digestive system that includes your stomach and intestines is referred to as the *gastrointestinal tract,* often abbreviated as the GI tract. Valves inside your heart regulate the blood flow between your heart's upper chambers, called atria, and its lower chambers, called ventricles. These valves are known as *atrioventricular valves,* which is often shortened to AV valves.

An **acronym** is technically a word formed from the first or key letters of each word in the full name. It is pronounced as if it is a word. For example, AIDS is the acronym for <u>A</u>cquired <u>I</u>mmune <u>D</u>eficiency <u>S</u>yndrome, and is pronounced like the word 'aids'. Sometimes regular abbreviations are formed like an acronym by using only certain letters from multiple words. For example, HIV stands for human immunodeficiency virus, the cause of AIDS, but we pronounce the acronym, AIDS, as a word, while spelling out the abbreviation – H-I-V. Many sources lump abbreviations and acronyms together, and certainly acronyms are a special type of abbreviation, but now you know the difference.

Abbreviations and acronyms can be problematic, though. SAD is an acronym for a condition called *seasonal affective disorder.* This disorder can be characterised as serious winter blues, apparently brought on by decreased daylight in winter months. The depression that accompanies it can be severe, making the acronym particularly suitable. However, a search of the medical abbreviations section of Medilexicon.com (see Web Resources at the end of the chapter) identified 21 terms that this abbreviation matches, and I found even more matches at other sites. They include:

- Separation anxiety disorder,

- Social anxiety disorder,

- Small airway disease,

- Systemic autoimmune diseases, and

- Sporadic Alzheimer's disease.

As you see, it is important to know the full name as well as the abbreviation, and to consider the context of how an abbreviation is

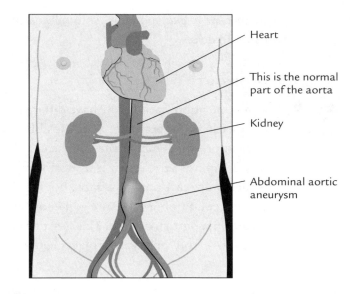

Heart

This is the normal part of the aorta

Kidney

Abdominal aortic aneurysm

FIGURE 3.2 An abdominal aortic aneurysm (Triple A or AAA).

used when determining its meaning. For example, an abbreviation used in medicine is AAA, referred to as a 'Triple A', which stands for abdominal aortic aneurysm (**Figure 3.2**), a condition often seen in cadavers in an anatomy lab. This is a weakening and ballooning of the aorta, the body's largest artery, where it ends in the abdomen. If it ruptures, it can kill you quickly. I could also be discussing my membership of the Amateur Athletic Association or be concerned about the security of internet shopping, when I would want to know about Authentication, Authorisation and Accounting systems.

Here are two last cautions about abbreviations: there may be more than one for the same thing, and some that sound quite similar may be different. For example, a *CAT* scan is an image created through a technology called computer-aided tomography, but these are now more commonly called *CT* scans. An *ECG* and an *EKG* are the same thing. Both abbreviations stand for electrocardiogram, a tracing of your heart's electrical activity – it doesn't matter if you use a 'C' or a 'K'. But if you change the middle letter to an 'E', you are referring to an electroencephalogram, which is a tracing of your brain's electrical activity. Use abbreviations with caution, and don't worry much about them now – like eponyms, you will learn them as you go and it is easier to do so.

LOOK OUT

The same abbreviation may have different meanings in different care settings: for example, AF is the abbreviation for atrial fibrillation which means that the heart is beating in an irregular way. In an obstetric unit it is a short form for amniotic fluid which is the liquid surrounding a baby, but the mother could also have a cardiac condition. Make sure you know what the abbreviation means in a particular area. This is especially important for students who move from one type of placement experience to another.

When using abbreviations and acronyms, always be sure that you know the full names and their meanings, and that you list the letters in the correct order. ▪

TIME TO TRY

Let's see how many of these abbreviations and acronyms you either know or can figure out. Match the following abbreviations to their full names.

1. _____ FBC

(a) Red blood cell

2. _____ ABC

(b) Ice, compression and elevation – how you treat an injury to a joint to minimise the damage

3. _____ RBC

(c) Full blood count – a lab test in which all the types of blood cells in a blood sample are counted

4. _____ ICE

(d) Airway, breathing, circulation – the order in which you assess a victim when administering first aid

These abbreviations probably gave you little trouble – you just had to match the letters in the name with the abbreviation. FBC means full blood count, ABC is the assessment used for administering first aid, RBC is a red blood cell, and ICE is the acronym for ice, compression, and elevation.

✔ **QUICK CHECK**

1. What makes an acronym different from a regular abbreviation?

_____

2. SIDS stands for Sudden Infant Death Syndrome. Is this an abbreviation or an acronym? _____

Answers: 1. Unlike a regular abbreviation, an acronym is a word formed by key letters from the multiple words that it represents. 2. SIDS is an acronym, pronounced as it is spelled.

Putting Down Roots and **Building Descriptive Terms**

Most anatomy and physiology terms are descriptive and are built from two or more of three basic parts:

- A **prefix,** at the beginning of the word,

- A **root,** which is the main focus of the word, and

- A **suffix,** at the end of the word.

New words may be made simply by changing these parts (**Figure 3.3**). Some time ago, when learning English, you learned about prefixes and suffixes and how to use them to change one word to another, and we use them all the time. You will be amazed how many word roots, prefixes and suffixes you already know.

A prefix is a short addition placed at the front of a word root. Consider the root *cycle*, which means circle or wheel. We can change the specific meaning of this root by adding prefixes as follows:

- uni*cycle* – has one wheel

- bi*cycle* – has two wheels

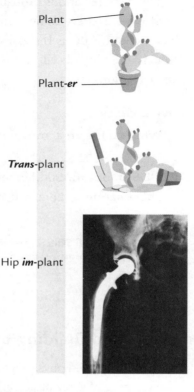

Plant

Plant-*er*

Trans-plant

Hip *im*-plant

FIGURE 3.3 Adding prefixes and suffixes to the root 'plant' will change the meaning.

- tri*cycle* – has three wheels

- motor*cycle* – has a motor

You probably knew the prefixes we used in our example. When you are reading anatomical terms, don't let the first glance worry you. Simply take a breath and look carefully at the word. Find the root, then examine any prefixes or suffixes that are used with it. Sometimes you may not know a particular word part at first, but you may work it out by thinking of words you *do* know that include that part. Let's try some.

TIME TO TRY

Here are some word parts and examples of words that use them. For each of the following, what do you think the word part means? Look at the examples and see if there is something they all share in common. This technique can help you determine the meaning of words with which you are unfamiliar – first look for the word parts and think of other terms that you do know that contain them.

Word part	Examples	Meaning
Root: cardi-	Cardiology, pericardium, cardiac arrest	_____ _____
Prefix: multi-	Multicellular, multitude, multilingual	_____ _____
Suffix: -ologist	Cardiologist, gynecologist, arthrologist	_____ _____

From this exercise, you probably can see that *cardi-* refers to the heart, *multi-* means many or multiple, and *-ologist* means one who is knowledgeable in or studies whatever the root says. *Cardiology* is the study of the heart, and a person who studies the heart is called a *cardiologist*. Note how similar these terms are, yet their meanings are different. The *pericardium* is the sac that surrounds the heart (*peri-* means *around*). And *cardiac arrest* refers to when the heart abruptly stops functioning, such as from a heart attack.

Let's consider the name of your current course: anatomy and physiology. The prefix *ana* means up or apart, and the root *tome* means a cutting. *Anatomy* is a combination of the two. It is the science that focuses on the parts and organisation of the body. But, when early folks tried to discover how we are built, all they saw was the outside. To discover how the body is organised, they did dissections – they *cut up* the bodies, *cutting away* outer structures to see deeper structures (**Figure 3.4**). See how descriptive the term really is?

Physiology comes from the root *physio*, which means nature or natural, and the suffix *–ology*, which means *the study of*, or *a branch of knowledge*. Physiology is the study of the branch of knowledge

FIGURE 3.4 Early anatomists explored the body's structures by cutting away the outer structures to see inside.

pertaining to nature. If you think of the natural sciences, which include chemistry and physics, this makes more sense – physiology is the study of how our body parts function, and that is largely determined by the laws of chemistry and physics.

Are you catching on to how easy this really is? Not only do descriptive terms convey a clear meaning, but they are easy to translate once you know the prefixes, suffixes and roots. You already know several of these, and you will learn plenty more as you go. To help jump start your language acquisition, **Table 3.1** lists several common prefixes and their meanings, **Table 3.2** lists several common roots, and **Table 3.3**

TABLE 3.1   Some common prefixes.

Prefix	Meaning	Prefix	Meaning
a-, an-	without	inter-	between
ab-	away from	intra-	within
ad-	to or toward	iso-	equal
alb-	white	later/o	side
ambi-	both	leuk/o	white
ana-	up, back again, throughout, against	macr/o	large
		mal-	bad
ante- or anter/o	before, in front of	medi-	middle
anti-	against	mega-	large
bi-/bis-	twice, double	melan/o	black
brady-	slow	micro-	small
circum-	around	mid-	middle
con-	with	mono-	one
contra-	against	multi-	many
cyan/o	blue	neo-	new
di-	two, twice, double	non-	not
dors/o	back	para-	along side of, beside
dys-	hard, difficult, bad		
ecto-	outside	per-	through, thorough, complete
en-	in		
endo-	within	peri-	around, near
epi-	on, upon	poly-	many
erythro-	red	poster/o	after, behind
eu-	well, good, easy	pre-	in front of, before
exo-	outside, outward	pro-	before, in front of
extra-	outside, beyond, in addition to	pseudo-	false
		quad-	four
hemi-	half	retro-	backward
heter/o	other	semi-	half
homo-	same	sub-	under, below
hydro-	water	super-, supra-	above
hyper-	over, above	syn-	with
hypo-	below, under	tachy-	fast
in-	in, into, on	trans-	across
infra-	below	tri-	three

TABLE 3.2 Some common word roots.

Root	Meaning	Root	Meaning
adeno	gland	gram	record
adipo	fat	graph	recording, writing
andr/o	male	gynec/o	female
angi/o	vessel	haem/o, -emia	blood
arter	artery	hepat/o	liver
arthr/o	joint	histo	tissue
aud or aur	hear	hydr/o	water
axilla	armpit	hyster/o, metr/o	uterus
brachi	arm	lapar/o	abdominal wall
bronch/o	lung air passageways	laryng/o	larynx, 'voice box'
burs/o	bursa, 'bag', (shock	lumb/o	lower back
	absorber between	mamm/o, mast/o	breast
	tendons and bones)	men/o	menstruation
cardi/o	heart	mening/o	membranes around
caud/o	tail		the brain and spinal
carp/o	wrist		cord
cephal/o	head	my/o, myos/o	muscle
cerebr/o	brain	nas/o	nose
cervic	neck	nephr/o, ren/o	kidney
chondr/o	cartilage	neur/o	nerve
clavi or cleido	clavicle	ocul/o	eye
colon/o	large intestine	oophor/o	ovary
cost/o	rib	op, opth	eye
cox/a	hip	orchid/o, test/o	testes (male gonad)
cubit	elbow	oste/o	bone
cyst/o	bladder, sac	oto	ear
cyt/o	cell	patho	disease
dent/o	teeth	pecto	chest
derm/o	skin	pes, ped, pod	foot
encephal/o	brain	phlebo or ven/o	veins
entero	intestine	plantar	sole of foot
gastr/o	stomach	pneumo/pulmo	lung
genesis	origin	procto	anus/rectum
gingivo	gums	pulmo/o	lung
glosso/linguo	tongue	ren	kidney
glute	buttocks	rhin/o	nose

TABLE 3.2   Some common word roots, continued.

Root	Meaning	Root	Meaning
salping/o, salpinx	uterine tube	tom	cut
scope	examine closely	trache/o	trachea, 'windpipe'
stasis	stay the same	ur/o, -uria	urine
stomato	mouth	vas/o	vessel, duct
talo	ankle	veno/phlebo	vein

TABLE 3.3   Some common suffixes.

Suffix	Meaning	Suffix	Meaning
-ac, -al	relating to	-ogen	precursor
-algia, -algesia	pain	-ologist	one who studies/ specialise in
-ase	enzyme		
-ate	do	-ology	the study of
-cide	kill	-oma	tumour (usually)
-cise	cut	-osis	full of
-cyte	cell	-ostomy	'mouth-cut'
-ectomy	removal of, cut out	-otomy	to cut into
-emia	condition of the blood	-pathy	disease of, suffering
-form	shaped like	-penia	lack
-genic	produced by	-plasty	surgical re-shaping
-gram	a recorded image	-plegia	paralysis
-graph/y	recording an image	-philia	affection for
-ia, -ic	relating to	-pnea	breathing
-ile, -illa, -illus	little version	-porosis	porous
-in	substance	-rrhage	excessive, abnormal flow
-ism	theory, characteristic of		
-itis	inflammation	-rrhea	discharge or flow
-ity	quality	-scopy, -scopic	to look, observe
-ium	thing	-sis	idea (makes a noun, typically abstract)
-ise	do		
-logy	study of, reasoning about	-stasis	to stop
		-stenosis	abnormal narrowing
-lysis	destruction, rupture	-tomy	cut
-megaly	large or enlarged	-ule	little version
-oid	resembling, image of	-um	thing (makes a noun)

TIME TO TRY

Now you can practise putting your new knowledge to use. Match the following terms with their meanings.

1. _____ arthritis   (a) Surgery through the abdominal wall

2. _____ otolaryngologist   (b) Surgical removal of the appendix

3. _____ appendectomy   (c) Difficult or painful menstruation

4. _____ dysmenorrhea   (d) Inflammation of the joints

5. _____ laparotomy   (e) Ear and throat doctor

Answers: 1. d., 2. e., 3. b., 4. c., 5. a.

lists several common suffixes. Don't worry about learning them all now – they are there to help you start building your language skills. You should read through the terms and see which ones you already recognise.

Mastering word roots, prefixes and suffixes is the most effective way to expand your A&P vocabulary because they can be recombined into countless new words. ■

How to Make a Jack-o'-Lantern: **Combining Word Roots**

When learning anatomical and medical terms, you should know some simple rules. You may have noticed in Table 3.2 that some roots end with '/o'. That means the root can be used with or without an 'o' added at the end. When the 'o' is added, the modified root is referred to as the **combining form**. It is awkward to pronounce a word formed from a root that ends in a consonant letter and another root or suffix that starts with a consonant. For these roots, a vowel is inserted in the middle to make it easier to pronounce. This is most often an 'o'. For example, inside your body your abdominal cavity and your pelvic cavity are continuous, so they are often referred to as one cavity by combining the two roots: *abdomin* and *pelvic*. The resulting word would be *abdominpelvic.* Instead, we insert an 'o' between the roots to get *abdominopelvic,* which rolls off the tongue more easily. I call this rule the jack-o'-lantern construction.

TIME TO TRY

Combine the following word parts, then define the term.

1. gastr   +   -scopy = _____
   (stomach) (close examination)
   What does this term mean? _____

   _____

2. rhin   +   -plasty = _____
   (nose) (surgical reshaping)
   What does this term mean? _____

   _____

3. haem   +   -rrhage = _____
   (blood) (excessive flow)
   What does this term mean? _____

   _____

4. encephal   +   -itis = _____
   (brain) (inflammation)
   What does this term mean? _____

   _____

5. cardi   +   myo   +   pathy = _____
   (heart) (muscle) (disease)
   What does this term mean? _____

   _____

Did you get those? The first two combine to form *gastroscopy,* which means examining the stomach by inserting a viewing instrument (gastroscope) into it. The next pair form *rhinoplasty,* which is surgical reshaping of the nose – a nose job! The third pair combine to form *haemorrhage,* which means excessive blood, and refers to blood loss (bleeding). The fourth pair form *encephalitis*, which is inflammation of the brain. Note that this pair does not require the 'o' for combining the two parts because the suffix begins with a vowel. Finally, you combined two roots and a suffix to form *cardiomyopathy*, which is disease of the muscles of the heart. Even though the root ends with a vowel, the 'o' is still added – always be on the lookout for exceptions.

Here is another rule to remember: *Spelling counts!* In some cases, changing a single letter can make a big difference in the meaning of the word. For example, two terms that we use to discuss movement of body parts are *abduction* and *adduction*. Note that the only difference between the two is the second letter. However, abduction means to move a body part out to the side, or away from the midline, but adduction means moving towards the midline. Only that one letter distinguishes these two opposite terms. Here is another pair: *ilium* and *ileum*. The ilium is part of your hip bone, but the ileum is the last part of your small intestine.

LOOK OUT

During or following surgery involving joints, you may be asked to keep the limb abducted or adducted. If you don't know the difference you may disrupt the progress of the patient's recovery.

Remember the importance of spelling – changing even one letter may change which structure you are naming. ■

Another manoeuvre that can be tricky with these terms is converting them from the singular form to the plural form. Some of the rules are the same as in English. If you look at one muscle and then another muscle, you are, indeed, examining two muscle**s**, and more than one bone are bone**s**. But there are some unique rules for pluralising many terms that you will encounter. More than one vertebra are vertebra**e**, and the small bone at the end of each of your fingers is a phalanx, but all of these bones in your fingers together are called phalang**es**. **Table 3.4** shows the rules to guide you with pluralisation.

TABLE 3.4 Basic rules for changing words in singular form to plural.

If the word ends in	Do this first	Then add	Examples
-a		Add -e	*vertebra* becomes *vertebrae*
-ax	Drop -ax	Add -aces	*thorax* becomes *thoraces*
-ex or -ix	Drop -ex or -ix	Add -ices	*cortex* becomes *cortices*
-ma		Add -ta	*dermatoma* becomes *dermatomata*
-is	Drop -is	Add -es	*anastomosis* becomes *anastomoses*
-nx	Change -x to -g	Add -es	*larynx* becomes *larynges*
-on	Drop -on	Add -ia	*ganglion* becomes *ganglia*
-us	Drop -us	Add -i	*nucleus* becomes *nuclei*
-um	Drop -um	Add -a	*ischium* becomes *ischia*
-y	Drop -y	Add -ies	*biopsy* becomes *biopsies*

Assume **The Anatomical Position**

Now that you understand how to read and form most of the terms used in anatomy and physiology, we can move on to some areas of specialised terminology that do not fit this pattern. The first of these to introduce is the **anatomical position**. This term refers to a specific body position that is used universally as a common reference point for the positions of body structures.

REALITY CHECK

If you are getting a bit tired of sitting, you might actually try the following positions. If you are feeling a little less active, just envision them.

1. You are lying on your couch or bed, face down.

What part of you is the top? _____

What is the bottom? _____

Which way is left or right? _____

Where is the front or the back? _____

2. You are stretched out in a recliner – upside down! Your feet are up over your head. Now what part of you is the top and what is the bottom? Next, sit in the chair the correct way. Do any of those positions change? _____

_____

You likely see from this exercise that our perceptions of up and down, right and left, and front and back may change with changing body position. That is rather inconvenient if we are trying to describe the locations of structures within the body. To avoid such confusion, we always assume that the body is in the anatomical position, whether it is or not. I could just describe this position, but you will understand it better if you try it out, so on your feet please (this one is easy)!

APPLYING THE THEORY

Placing a patient in the correct position for examinations, treatments or to relieve symptoms such as breathlessness or pain can make a difference to patient comfort and well-being. Using the correct terminology means that everyone understands what is required. If you are asked to place a patient in the left lateral position with the right knee flexed, would you know how to explain to the patient what they needed to do?

TIME TO TRY

Stand straight and look directly forward. Be sure the soles of your feet are planted firmly on the ground with your toes pointing forward. Let your arms hang straight down at your sides. In this natural position, your palms are probably facing your thighs. To move into the anatomical position, all you need to do is turn your palms so they are facing forward. Note this position (**Figure 3.5**) – you are now in the anatomical position! Remember it, because that is always the starting point when discussing anatomical relationships.

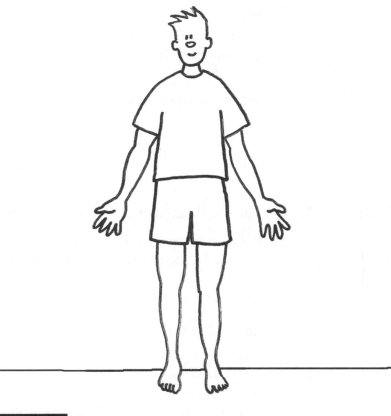

FIGURE 3.5   The anatomical position.

 All anatomical structures are named based on the anatomical position. ▨

Not only does the anatomical position provide a common reference point, it also provides specific benefits. In this position, it is obvious that the limbs are parallel to the body and everything is straight. You do not see from the outside that there are two bones in the forearm – the *radius* and the *ulna*. When your palms are facing in toward your thighs as you usually hold them, these two bones are actually crossed, but in the anatomical position, they are parallel to each other. In this position, all the muscles, vessels and nerves in the forearm are also running a straight course.

✔ QUICK CHECK

1. In what position are the hands when the body is in the anatomical position? _____

_____

2. Why is the anatomical position used? _____

_____

Anatomy ABC's: **Basic Anatomical Terminology**

Now that you know the anatomical position, we can introduce some terms that are used when discussing body structures. You will learn these, and plenty more, in class, but students sometimes struggle with these terms. You will have an advantage if you get started on them now.

Relative Position Terms

The following terms are often called relative position terms because they are used to compare the locations of structures. Many of these terms are paired, and that is the best way to learn them. When learning these, be sure to remember always to assume that the body is in the anatomical position.

Superior and **inferior**: *Superior* and *inferior* obviously mean above (toward the top) and below (toward the bottom), respectively. They have all your life, so you already know these terms. For example, your head is superior to your neck, and your thorax (chest) is inferior to your neck.

Cephalic and **caudal**: Look up these roots in Table 3.3 on page 109. *Cephalo-* means _____ and *caudo-* means _____, so these two terms mean toward the head or toward the tail. They are sometimes used in place of superior and inferior, which usually works fine. Just don't use caudal to refer to structures located below the tailbone.

Anterior and **posterior**: I am pretty sure you know these, too. They simply mean towards the front or towards the back. Your sternum is anterior to your lungs, for example, and your spine is posterior to them.

Ventral and **dorsal**: *Ventral* refers to the belly, and you already know dorsal – what is that thing on a shark that you see sticking up out of the water? _____ fin. Yes, that is the dorsal fin, and it is on the shark's back. *Dorsal* means toward the back. For example, your umbilicus (belly button) is ventral, and your spinal cord is dorsal. Most animals get around on all four legs, so for them these terms are synonymous with inferior and superior – that fin is on the top of a swimming shark, for example. Because we walk upright, our bellies are toward our fronts and our backs are, of course, toward our backs. So for us, *ventral* means the same as *anterior* and *dorsal* means the same as *posterior*.

Medial and **lateral**: Medial and lateral refer to positions along a side-to-side axis. Imagine an axis passing directly down through the centre of your body all the way from the top of your head to the floor. *Medial* means towards the midline of the body, and *lateral* means to the side, away from the middle. A special term, **median**, refers to being exactly on the midline. Your heart is medial to your lungs, for example, and your arms are lateral to your trunk. Your trachea and oesophagus are both in a median position.

Superficial and **deep**: You should already understand these as well. *Superficial* means closer to the surface, and *deep* means closer to the centre of the body. In your head, for example, your skin is superficial and your brain is deep.

Proximal and **distal**: These terms are usually the most challenging for students, but they are actually rather simple. *Proximal* means closer to and *distal* means farther away. The question, though, becomes closer to or farther away from *what*? The reference point is some point of origin. This is easy for extremities – the point of reference is where the limb begins – where the limb arises from the trunk. So, in the upper extremity, the elbow is distal to the shoulder, but it is proximal to the wrist. Simply think of where the starting point is – here it is the attachment at the shoulder. Then visualise running your finger along the limb, starting where the limb starts. If you are comparing two locations, whichever you touch first is proximal to the other one. Thus, for the upper extremity the beginning is at the shoulder. Tracing out from there, the elbow is distal, or farther away from the starting point. You reach the elbow before you reach the wrist, so the elbow is *proximal* to the wrist – the elbow is closer to the beginning of the limb than the wrist is.

WHY SHOULD I CARE?

During your class, you will examine many structures whose names include relative position terms. When you have a test or examination, you may be asked to label a diagram, so you need to know these names. For example, you have both a femoral artery and a deep femoral artery in your thigh. If you leave out the word 'deep', you are incorrectly referring to an entirely different blood vessel. Similarly, the anterior and posterior tibial arteries supply different areas of the leg, as their names imply.

✔ QUICK CHECK

1. Your ankle is in what position compared to your toes? _____

2. Your nose is in what position on your face? _____

Answers: 1. proximal, 2. median.

APPLYING THE THEORY

When a patient has poor circulation to their legs, you may be asked to assess the pulses in the femoral, popliteal, anterior and posterior tibial arteries or the dorsalis pedis. You need to know where they lie in order to be able to feel them. Other patients may have plaster of Paris applied to stabilise a fracture. Tight bandages, plaster casts and elastic stockings can impair the circulation to a limb and one of the early warning signs of a problem may be a reduced pulse to the area, so you need to know which blood vessel to palpate.

Planes and Sections

We discussed previously that anatomy was originally revealed through dissections, which meant cutting into the body, removing parts and cutting into the parts as well. Looking at structures from different angles gives us a better understanding of their true natures, but we then need words to help distinguish between these different views.

You can think of a **plane** as resembling a line that passes completely through the body in a particular direction, all the way from top to bottom or from side to side, and a **section** is a cut made along a certain plane. In class, you will discuss body parts that lie on certain planes within the body, and you will examine models and specimens that are cut in certain sections. Let's explore some of them now.

There are three major planes that pass through the body:

■ sagittal,

■ coronal or frontal, and

■ transverse or horizontal.

A **sagittal plane** passes vertically through the body from front to back, as shown in **Figure 3.6a**. You can think of it as dividing the body into right and left parts that may or may not be equal. If the plane passes directly through the midline of the body, it is called a **midsagittal plane** and it produces equal right and left parts. Recall, though, that a term that refers to the exact midline is *median*, so this same plane is also correctly called a **median sagittal plane**, or sometimes simply a **median plane**.

FIGURE 3.6   Major planes through the body include **(a)** sagittal, **(b)** coronal, and **(c)** transverse.

A cut made along this plane could be called a midsagittal section, a median sagittal section, or simply a median section. You will likely see this type of section in the lab when you examine anatomical models such as the head or the pelvis that have been cut along this plane.

The second plane is the **coronal plane**, shown in **Figure 3.6b**. This plane also passes vertically through the body, but from side to side, so it gives you front and back parts. In fact, this plane is also known as a **frontal plane**. Your lecturer may have a preference for which term you use, but I suggest to my students that they use frontal plane simply because it is easier to remember.

The third and final major plane is the **transverse plane** (**Figure 3.6c**). This is the only **horizontal plane**, and it is also referred to by that name, which is the one I prefer, again because it is easier to remember. This plane divides the body into top and bottom parts.

PICTURE THIS

1. Imagine you are at a magic show. Suddenly the magician pulls you out of the crowd and places you lengthwise into a large box. (If you can, please lie in that position now.) Much to your horror, he begins to saw you in half at the waist!

   Into what kind of parts are you separating? _____

   _____

   Which plane does this cut represent? Don't forget the anatomical position! Stand upright if it helps you get the answer. _____

   _____

2. Now, please stand in the anatomical position. Imagine there is a zipper running down the midline of your body from front to back, passing between your eyes, straight through your nose and on down through your chin and neck. Now start pulling down the zipper.

   Into what kind of parts is your body separating? _____

   _____

   Which plane does this represent? _____

3. Again start from the anatomical position. You may have seen circus clowns act like they are running a string in one ear and out the other. Visualise that, but change the string to a thin wire and start at the top of the head. Imagine doing that same side-to-side movement all the way down your body.

Into what kind of parts is your body separating? _____

_____

What kind of plane does this represent? _____

In the first part of this exercise, you should see that if the box is stood upright so that you are in the anatomical position, the magician's cut divides you into top and bottom sections and represents the transverse or horizontal plane. The zipper down your midline splits you into right and left parts and represents a sagittal plane – technically midsagittal, because it is on the midline. The final scenario splits you into front and back parts, representing a coronal, or frontal, plane.

Sometimes, though, we are looking not at the whole body, but rather at individual parts that have been cut open. Because they are cut, the particular view is referred to as a section. **Sagittal sections** are common through the head and trunk, but are less obvious in the extremities or in individual organs. For cuts through those, we more often use three different terms:

■ longitudinal,

■ cross, and

■ oblique.

A **longitudinal section** is cut along the *long* axis (**Figure 3.7a**). A **cross-section** is cut directly *across* the long axis, and is also called a **transverse section** (**Figure 3.7b**). An **oblique section** is a section cut along any other angle (**Figure 3.7c**). I tell my students to associate the 'o' in oblique with either the term 'off' or 'odd' because an oblique angle is off of the main axes, making it the odd angle.

(a) Longitudinal section   **(b)** Horizontal or cross-section   **(c)** Oblique section

FIGURE 3.7   Sections include **(a)** longitudinal, **(b)** transverse or cross, and **(c)** oblique.

TIME TO TRY

1. Obtain a long object through which you can cut, such as a carrot, a chocolate bar, or a drinking straw, and some scissors or a knife. (Be careful – sectioning yourself is not part of this exercise!)

2. Hold the object upright and visualise the transverse plane. Now lay it down and cut along that plane about midway down the object. What kind of section did you just make? _____

3. Keeping the object in that position, cut one of the halves straight down the midline.

   What kind of section did you just make? _____

4. Take the remaining half and cut it at any angle different from the first two.

   What kind of section did you just make? _____

When you cut through the object in a transverse plane, you are making a cross-section. Cutting these halves down the midline makes longitudinal sections. Finally, cutting the object at any other angle produces oblique sections.

APPLYING THE THEORY

Many of your patients will undergo a special type of X-ray, called a computerised axial tomography scan or CAT scan, sometimes abbreviated to CT scan. These scans take pictures in cross-section slices which can give a three-dimensional picture. (Look up the word tomography on pages 108–109.)

✔ **QUICK CHECK**

1. What are three names for a plane passing from front to back and directly through the midline of the body? _____

_____

2. A blood vessel is a tube-like structure. To remove a small section of a vessel, perhaps for a coronary bypass, you need to make two cuts. What kind of sections will you likely use? _____

_____

Answers: 1. Midsagittal, median sagittal, median. 2. Cross-sections.

In this chapter, I threw a lot of new terms at you. Most chapters in your textbook will do the same. At times, you may think that learning all the terms is impossible. Just when you think you know the name of a structure, you might learn that it has another name as well. Think about the red blood cell that carries oxygen in your blood. Do you know any other names for a red blood cell? You might – it is also called an **erythrocyte**, or it goes by the abbreviation RBC.

Multiple names are not uncommon in anatomy and physiology. In fact, my students often hear me say 'Why name it once if you can name it two or three times and confuse people?' At times, the confusion may seem intentional, but realise that there are many ways to name things – formal names (erythrocyte) and informal names (red blood cell), for example. Mastering the language can be challenging, but you are now armed with important tools to help you. You will also find that many terms are defined in your textbook, either in the chapters or in the glossary. And you can always purchase a medical dictionary or refer to an online version – they are invaluable. If your schedule allows, you might consider also taking a medical terminology course while taking A&P; just don't overload yourself.

APPLYING THE THEORY

It won't be all new all of the time! There are some other tricks you can try to help to remember the language. If you learn the names of the bones and major organs you will just need to adapt the nouns to adjectives to remember many of the other structures. If you are asked to feel the radial pulse of a patient, even if you have not heard the term before but you know where the bone called the radius is in the arm, then you can make an educated guess that you are being asked to feel the pulse near the wrist on the thumb side of the arm.

But remember the traveller who helped open our chapter. You now know to start your journey through each chapter in your textbook by learning the language. Know the terms so you can understand the material. Don't just memorise words – look up their meanings, look at how they are built, make and use flash cards, and be able to use your new words. When you see or hear a new word think about its structure and make an educated guess. Thinking about the words makes them easier to remember. Don't get discouraged if you don't master all the terms in this chapter – those tables are long! Get in the habit of talking anatomy and physiology with your study group, friends and family, for example. Learn what you can now and the rest will come as you go.

People often move to foreign-speaking countries and learn the language as they go. They learn by experience and practice. Students have been surviving in the foreign-sounding world of A&P for ages as well, and you have already had a crash course in the language. Enjoy your adventure!

FINAL STRETCH!

Now that you have finished reading this chapter, it is time to stretch your brain a bit and check how much you have learned.

RUNNING WORDS

At the end of each chapter, be sure you have learned the language. Here are the terms introduced in this chapter with which you should be familiar. Write them in a notebook and define them in your own words, then go back through the chapter to check your meaning, correcting as needed. Also try to list examples when appropriate.

Descriptive term	Anterior	Sagittal plane
Eponym	Posterior	Midsagittal plane
Abbreviation	Ventral	Median sagittal plane
Acronym	Dorsal	Median plane
Prefix	Medial	Coronal plane
Root	Lateral	Frontal plane
Suffix	Median	Transverse plane
Combining form	Superficial	Horizontal plane
Anatomical position	Deep	Sagittal sections
Superior	Proximal	Longitudinal section
Inferior	Distal	Cross-section
Cephalic	Plane	Transverse section
Caudal	Section	Oblique section

WHAT DID YOU LEARN?

Try these exercises from memory first, then go back and check your answers, looking up any items that you want to review. Answers to these questions are at the end of the book.

PART A: USING TABLES 3.1 TO 3.3, MATCH THE FOLLOWING TERMS WITH THEIR DESCRIPTIONS

1. _____ leukocyte

2. _____ endocarditis

3. _____ subclavian artery

4. _____ antecubital

5. _____ adipocyte

(a) Paralysis of all four limbs

(b) An image of the breast

(c) An enlarged liver

(d) Surgical removal of the uterus

(e) A white blood cell

continued overleaf

6. _____ mammography

(f) Inflammation of the lining inside the heart

7. _____ mammogram

(g) A fat cell

8. _____ hysterectomy

(h) An imaging technique used to assess the breasts

9. _____ hepatomegaly

(i) An artery located under the clavicle

10. _____ quadriplegia

(j) The area in front of the elbow

PART B: ANSWER THESE QUESTIONS

1. In what way is the anatomical position different from how you normally stand?

_____

_____

_____

2. Which type of plane divides the body into right and left portions?

_____

3. Which of the following is an eponym?

_____

(a) CPR   (b) SIDS
(c) cardiac sphincter   (d) McBurney's point

4. Which of the following is an acronym?

_____

(a) CPR   (b) SIDS
(c) cardiac sphincter   (d) McBurney's point

5. Which two planes pass vertically through the body?

_____

_____

6. Using Tables 3.1 to 3.3, construct a descriptive term for each of the following:

   A condition in which there are (too) many cells in the blood. _____

   A condition in which the liver is inflamed. _____

   _____

7. Using the tables, define each of the following terms:

   arterial stenosis _____

   _____

   chondrocyte _____

   _____

8. Provide the plural form of the following terms:

   pharynx _____

   mitochondrion (part of a cell)

   _____

   coxa (hip bone) _____

9. What is the difference between *medial* and *median*? _____

   _____

   _____

10. You decide to trim your fingernails into points by cutting at an angle from each side to the centre. What kind of section are you making? _____

    _____

WEB RESOURCES

Here are some additional online resources for you.

■ *Medline Plus*

http://medlineplus.gov

This comprehensive website provides both a medical dictionary and a medical encyclopedia, as well as numerous links to other valuable sites.

■ *Medilexicon*

http://www.medilexicon.com/

This site has a wealth of information, including a medical dictionary and extensive listing of medical abbreviations, as well as a medical news section and links to other valuable sites.

■ *Voycabulary*

http://www.voycabulary.com/

This terrific site lets you open any web page through their program and turns all words in the web page into links to a dictionary. You just enter the web address in Voycabulary, and when the page opens, click on any word and you will get its dictionary definition.

■ *Who Named It?*

http://www.whonamedit.com/

This is an extensive listing of medical eponyms.

■ *Word Info*

http://wordinfo.info/words/index/info/list/A

This site offers an extensive alphabetised list of word roots, prefixes and suffixes and provides their meanings and origins.

A small clinical dictionary is useful to keep in your pocket, but don't be fooled by those describing themselves as pocket dictionaries. Some publishers must think your pocket is the size of a backpack!

4 Body Basics
The Hip Bone's Connected to the . . .

When you have completed this chapter, you should be able to:

- Understand the Biological Hierarchy of Organisation.
- Understand the relationship between anatomy and physiology.
- Understand basic principles of biology that govern body functions.
- Explain homeostasis and its importance to body functions.
- Differentiate between tissues, organs and organ systems.
- List major organs in each organ system and understand each system's general functions.

YOUR STARTING POINT

Answer the following questions to assess your knowledge.

1. The study of tissues is called _____.

2. What is meant by 'Form fits function?' _____

_____

3. Where does your body get the energy it uses for work? _____

_____

4. According to biological rules, what is the main reason for our existence? _____

5. How many categories of tissue are there? _____

6. Which organ system secretes hormones? _____

Climbing the Ladder: **The Biological Hierarchy of Organisation**

Now that you are armed with some new language skills, let's peek at what lies ahead in your studies of anatomy and physiology. You have a lot to learn along the way, but your journey is, literally, one of self-discovery – you are learning about yourself. That should make your adventures in A&P quite exciting! This chapter will introduce some basic concepts and principles to guide your path. Let's begin with how your course will be organised.

In science, we like to categorise. One classification system is known as the biological hierarchy of organisation. This hierarchy moves from the simplest level of structural organisation up to the most complex. Several levels of this system are relevant to the disciplines of anatomy and physiology, and these are shown in **Figure 4.1**. Your textbook and your course will follow this pattern. These levels, from the simplest to the most complex, are the atom, molecule, macromolecule, organelle, cell, tissue, organ, organ system and organism.

Answers: 1. Histology. 2. A part's structure reflects the job it does. 3. From food. 4. To survive and reproduce to continue the species. 5. Four. 6. Endocrine.

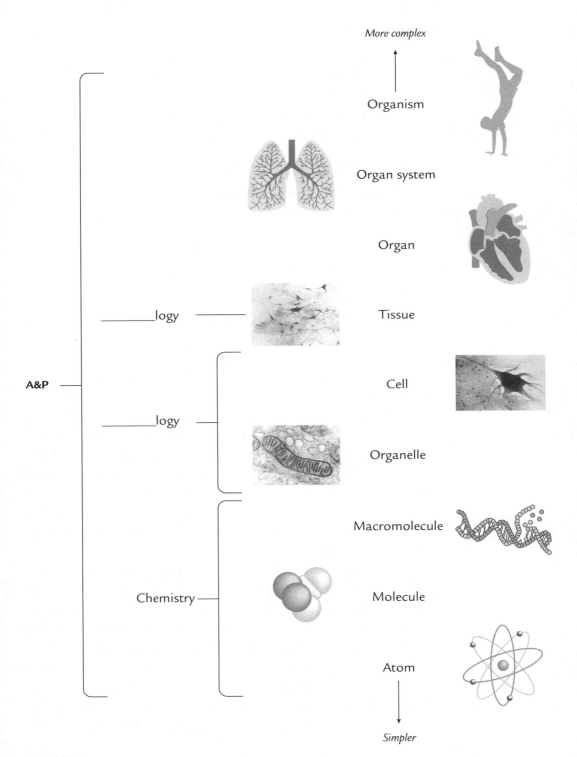

TIME TO TRY

Examine Figure 4.1. Note that the first three levels are part of the science of chemistry. The next two levels – organelle and cell – are covered by another science discipline. Look back at Tables 3.1 and 3.2 and determine the name of this science that studies cells and their structures. What is it? _____

Now, write that name in the appropriate line in Figure 4.1.

The next level is the tissue. Again, refer to Chapter 3 and determine the name of this branch of science.

What is it? _____.

Now add that label to Figure 4.1.

As you see from the illustration, the first three levels are **chemistry**, which we will review in Chapter 5. The next two levels are **cytology**, the study of cells, which we will explore in Chapter 6, and **histology**, the study of tissues. To study both cytology and histology requires magnification, typically through a microscope, so these disciplines are sometimes referred to as **microscopic anatomy**. The more complex levels – organ, organ system and organism – will comprise the bulk of your anatomy and physiology course. Structures at these levels are visible without magnification, so their study is also called **gross anatomy** – not gross meaning *horrible*, but rather gross meaning *large*.

All matter is made of **chemical elements**. The smallest piece of an element is an **atom**, and atoms can unite to form **molecules**. To understand this, consider water, arguably the body's most important nutrient. One atom of the element oxygen combines with two atoms of hydrogen to form a molecule of water (H_2O). Water molecules are quite small, but some molecules, such as fat, DNA, starch and proteins, are rather large, so they are referred to as **macromolecules**. Atoms, molecules, and macromolecules provide the nutrients and building materials our bodies need to stay alive and healthy, and they participate in chemical reactions that do all of the work performed in our bodies.

Macromolecules can unite to form complex structures called **organelles** that carry out functions inside cells. Organelles include cellular structures like the nucleus or a mitochondrion. **Cells** contain the combination of organelles necessary to sustain life, so cells are the first level of organisation that we consider to be alive. In fact, cells are the basic units of all living organisms. **Tissues** are groups of cells organised to perform some common function. For example, muscle tissue contracts to provide movement. Tissues then organise into larger functional units called **organs**, such as your lungs, heart, and liver. Each organ performs at least one specific job. Multiple organs combine to form **organ systems**, each with some overall function. For example, the organs in your cardiovascular system are your heart and your blood vessels. The heart is the pump and the vessels are the pipes, so to speak, through which your blood travels to deliver nutrients and oxygen to your cells and to haul away their wastes. Collectively, your organ systems do the work needed to keep you – the **organism** – alive!

PICTURE THIS

You decide to build a house. You use wood that came from trees, which were once living organisms. Explain how the first five levels of the Biological Hierarchy apply to the tree from which your wood came. _____

_____

_____

_____

Trees are made of cells containing organelles that allow the tree to grow and form woody material, such as plant fibre, which is a macromolecule. Macromolecules are built from smaller molecules and they, in turn, are built from atoms.

Now let's use the idea of a house as an analogy for *you*, a living organism, to trace the structural progression from simple to complex. Various building materials (cells) are used to form the walls, floors and ceilings (tissues). You add various systems such as plumbing to move the water in and your waste out, electricity to provide energy, and a heating and cooling system to regulate the air. The latter has a boiler, an air conditioner and ducts. Each of these items has a unique job, so they are like organs. These 'organs' form a complete system (the organ system). Once you get all these different 'systems' in place, you have a fully functional house (the organism).

✔ QUICK CHECK

Fill in the missing levels from the Biological Hierarchy of Organisation.

1. atom, molecule, _____macromolecule_____, organelle

2. cell, __tissue__, organ, __organ system__, organism

Answers: 1. macromolecule. 2. tissue; organ system.

Some Things Never Change: **Basic Biological Rules**

I find most of anatomy and physiology to be beautifully simple and understandable, and I hope you will soon see what I mean. At first you may be tempted to just memorise the names and locations of anatomical structures, but you must move beyond memorisation to truly understand the processes involved in the physiology. When studying the course material, always look for the logic and reasoning behind it. Your task will be easier if you understand some basic rules that govern the human body. These rules will help you understand why the body does what it does. There are, of course, exceptions to most rules, but these concepts should still help guide your thinking as you try to understand the 'whys' behind your learning.

Life Begins at the Cell

Humans are made of trillions of cells, each of which is, or at least started as, a living unit. Many of our body functions occur within our individual cells. Our cells also have highly specialised functions and they communicate with each other, so many body processes also occur through the coordinated action of groups of cells. In fact, all of our organ systems are interrelated and work together to keep us alive. Always keep in mind that all body processes begin at the level of the cell.

Form Fits Function

Recall that anatomy is the study of the body's structure and physiology is the study of the body's functions. Although these sound different, they are closely related. You need to understand the parts and how they are put together to know how they work. Similarly, if you know what a body structure does, you can usually predict how it is organised to do its job. A common theme throughout biology is that 'Form fits function'. This means that all body parts have specific structures that allow them to perform their jobs most efficiently. A part's shape and organisation reflect what it does, and similarly the job that is needed affects the structure that a part will have.

Let's consider the heart (**Figure 4.2**). The heart's job is to circulate your blood, and it does this by constantly pumping your blood along its way. To accomplish this, the heart has receiving chambers, the atria, and sending chambers, the ventricles. Some of the blood returning to the heart comes from the lungs carrying a rich supply of oxygen, and is ready to head out through your body to supply all your cells. The rest of the blood returning to your heart has just been to your cells and dropped off much of its oxygen for their use, so it needs to go to the lungs to get more. Your heart is divided into right and left sides. This allows your blood to travel through two different paths – the right side

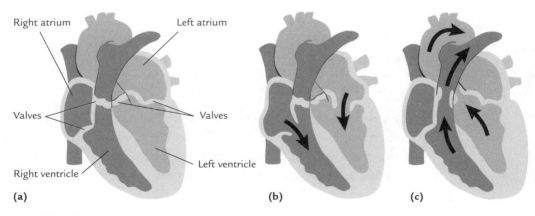

FIGURE 4.2 The anatomy of the heart. **(a)** The heart chambers and valves. **(b)** As the ventricles fill, the valves at the entrances to the ventricles are open but the valves at the exits are closed. **(c)** As the ventricles empty, the valves at their entrances are closed but the valves at their exits are open.

sends the deoxygenated blood out to the lungs, and the left side sends the freshly oxygenated blood out to your body's cells.

Your blood circulates best if it is always moving in a single direction – forward. Doorways called valves are part of the heart's anatomy. One set allows blood to enter your ventricles and the other set allows it to leave these chambers. To fill the ventricles, the doors into the ventricles are open but the exit doors are closed (**Figure 4.2b**). When the ventricles contract, the entrance doors close so the blood must move forward, pushing the exit doors open so it can leave (**Figure 4.2c**). The two sides of the heart keep the oxygenated and deoxygenated blood from mixing, ensuring that the tissues receive blood with the highest level of oxygen. Clearly the heart's anatomy is uniquely suited to its function. In other words, the heart's form fits its function.

APPLYING THE THEORY

It is not uncommon for one of the valves in the heart to become faulty, either not opening fully or closing properly. The anatomy (structure) is no longer suited to its function (as a pump) and so problems arise because blood can't get out quickly enough when the heart beats, or blood leaks through a valve that should be closed. Listen to a heart beating using a stethoscope. The sounds you hear are made by the valves closing.

TIME TO TRY

Examine **Figure 4.3** to answer the following questions.

1. Figures 4.3 (a) and (b) show thin and thick skin. Thin skin is found in most areas of our body, but the soles of your feet have much thicker skin. How do you think the thickness of the skin on the soles of your feet is related to the function of your feet?

2. Let's try the opposite approach. You use your hands to do all kinds of work, some of it physically hard. The palms of your hands experience tremendous wear and tear. What type of skin do you think covers your palms? _____

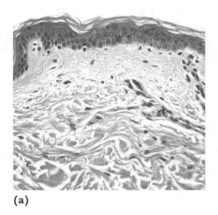

(a)

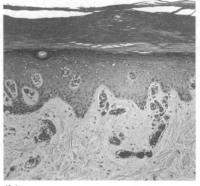

(b)

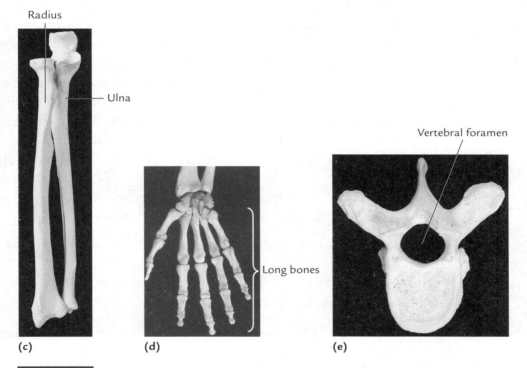

(c)   (d)   (e)

Radius

Ulna

Long bones

Vertebral foramen

FIGURE 4.3   **Form fits function. (a)** Thin skin covers most of the body's exposed areas. **(b)** Thick skin protects the soles of your feet and the palms of your hands. **(c)** Long bones in the forearm. **(d)** Long bones in the hand. **(e)** A thoracic (upper back) vertebra from the spinal column.

3. Figures 4.3 (c) and (d) both show long bones, but the long bones in your hand are much shorter than the long bones of your forearm. How does the function of these two areas explain this structural difference in their long bones?

4. Figure 4.3 (e) shows a top view of a thoracic vertebra from the vertebral (spinal) column. The vertebral foramen, shown here, is found in all vertebra. Why do you think that hole is there?

From the previous exercise, you should realise that thick skin is located in areas that get the most use, pressure and wear and tear. Thicker skin provides better protection for the underlying structures, and because your palms are used so extensively, they are covered by thick skin, like their counterparts in the lower extremity – the soles of your feet. Your forearm contains only two long bones, but there are 19 much shorter 'long' bones in your hand. This allows the hand to move into many different shapes and to grip objects tightly, making it one of our most versatile tools. Vertebral foramina are found in all of your vertebrae, stacked on top each other to provide a hollow and protected passageway for your spinal cord and the nerves arising from it.

You already know many of your body's structures and what they do. As you learn the rest, always ask yourself the following questions:

■ How is this structure built?

■ What is its job and how is it performed?

■ How do structure and function fit together?

When you can answer those questions, you are truly learning anatomy and physiology. When you examine a structure, the way it is built should give you clues about what it does. Similarly, if you know what a structure does, you should be better able to understand how and why it is built the way it is.

LOOK OUT

People who are ill or frail tend not to be very mobile. Staying in one position for a long time puts pressure on the skin and blood vessels that are not built to withstand this extra wear and tear. Make sure that you help them to change position frequently so that the delicate skin does not break down from pressure damage.

✔ **QUICK CHECK**

What is meant by 'Form fits function?' _____

_____

Answer: A part is organised in a manner that makes it most suited to the job it needs to perform. Its structure reflects its function.

 A body part's structure reflects its function. ▪

Life Requires Energy

All living organisms use energy. **Energy** is defined as the ability to do work, so it is required to do all of your body's work, or **metabolism** – virtually everything that is occurring in your body at any given time. Almost all energy on Earth ultimately comes from the Sun and is first harnessed by plants through photosynthesis. During this process, plants take in carbon dioxide and water from the environment and use solar energy to convert these molecules into chemical energy, stored in sugars. Plants also release oxygen as a waste product.

We cannot harness energy from the Sun, so we eat the plants, or the animals that ate them, to get the chemical energy they created and stored (Thank you, plants!). Our cells then use that stored energy to perform our normal body processes. The process by which we convert stored energy into a usable form is called **cellular respiration**, and we get more energy out of our food if we use oxygen in this process. Body parts that are the most active (do the most work) have the highest energy demand, and we must continually supply them with energy in order to do their work.

REALITY CHECK

Based on your current knowledge of your organ systems, which two organ systems do you think help us obtain and use energy from food molecules? _____

_____

Answer: The digestive and respiratory systems.

Conservation

Without energy, we die, so our bodies are very careful about how we use our energy. Our cells conserve energy whenever possible. During the day, when you are awake and active, you use a large amount of energy to meet all your needs. When you are sleeping, many of your organ systems don't have to work as hard, and you use less energy. If you find yourself in a potentially dangerous situation, your body uses more energy to prepare you to respond, perhaps by fleeing from the threat. Once you are out of harm's way, your energy use again drops. In almost any situation, the body uses the least amount of energy it can to do its tasks.

LOOK OUT

When people are ill, elderly or have undergone surgery, they may need extra energy to repair the damage and become mobile. It is often a time when they feel least like eating so it is important for the staff caring for them to ensure that they receive food that is energy rich, appetising and is easy to eat.

PICTURE THIS

Our energy comes from the food we eat. In Système Internationale (SI) units, energy is measured in joules (J). This is a very small amount of energy, so we usually think about 1000 J, which is called a kiloJoule (kJ). You are probably more familiar with Calories, more correctly called kilocalories (kcal). You will notice the upper case C for Calorie – this is the same value as 1 kcal but is not part of the Système Internationale: 1 kcal is approximately 4 kJ. Food packaging usually has both kcal and kJ values of the contents on the wrapper. Think about energy as being body 'money' – if you do not use all your income you will accumulate savings. We have some control over our body's savings account by balancing energy intake with energy output. In this way we do not gain or lose weight and can maintain our normal body size.

1. How? _____

2. If we do not take in enough energy 'money', we cannot meet all of our energy demands, or, in effect, we cannot pay our bills. What happens in your home if you don't pay your power or your water bill? _____

3. What would happen to your organs if you did not give them enough energy? _____

4. If we earn more real money than we spend, what do we wisely do with the extra? _____

5. What do you think the body does with extra energy 'money' it takes in? _____

Most extra energy, again meaning kJ, is stored in the body as fat, regardless of what we eat. This is the body's version of saving for a rainy day, and we gain weight. As with our bank accounts, to lose stored income (weight), we need to spend more than we are taking in. That may be, unfortunately, very easy to do with real money, but your body 'bank' doesn't make it so easy to take those extra kJ out of storage because it is designed to conserve energy. It gladly lets you make deposits, but tries to keep you from making withdrawals. In fact, if you cut back dramatically on your caloric intake, your body becomes even more conservative and spends less energy, making it even harder to burn off those extra pounds.

JUST FOR FUN

Check out the nutrition label on your favourite food.

1. First, notice the serving size. Is that the amount you usually consume as a 'serving?' _____

2. How many kilocalories are there in a single serving? _____

3. What is this in kiloJoules (kJ)? _____

4. What is the recommended daily energy intake for

   (a) woman _____ kJ _____ kcal

   (b) man _____ kJ _____ kcal

Do you like to think of your energy intake, in kiloJoules or kilocalories?

Obtaining all the materials we need to make and maintain our body parts requires considerable energy. To minimise this cost, our bodies recycle. Many molecules are broken down to atoms and then built back into other molecules, over and over again. Think about a child's building block set. Children can spend hours building things, taking them apart, then building new ones – a car is stripped down, then becomes a castle, which is stripped down and rebuilt into a robot. Like the child, the body reuses both energy and materials.

I tell my students that the body is lazy – it prefers not to do anything it does not absolutely have to do. I further explain that, more accurately, the body is highly conservative – it doesn't make or maintain parts it doesn't need, and it spends as little energy as it can.

This theme of efficiency will help you understand many aspects of physiology. For example, the saying 'Use it or lose it' is physiologically quite true. Tissues respond to repeated stress by building up and becoming stronger. Aerobic exercise stresses your heart, which responds by becoming stronger and more efficient. Strength training stresses your skeletal muscles, which respond by becoming bigger, stronger and better at generating force. Weight-bearing exercise stresses the bones, so they thicken to become stronger. This is an important way for women to avoid the devastating consequences of osteoporosis (**Figure 4.4a**). But if you stop exercising and stressing these tissues, all

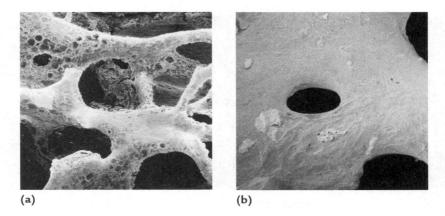

(a)   (b)

FIGURE 4.4   **(a)** Bone that is not stressed during the first two decades of life is more likely to develop serious osteoporosis in later years. **(b)** Healthy bone is dense and well developed and better able to resist age-related bone loss.
Source: (a) © P Motta and (b) © Steve Gschmeissner/SPL. Photo Researchers Inc.

the extra growth and development are quickly lost. Astronauts returning to Earth after being in a weightless state, for example, have lost measurable muscle and bone mass, leaving them slightly weakened upon their return. It takes energy and materials to maintain tissues, and the body's conservative nature prevents it from maintaining any that are not needed.

APPLYING THE THEORY

Sam Freeman is aged 27. He has been brought to the accident and emergency department with a suspected fractured tibia following a tackle while playing rugby.

Mrs Ada Green, aged 80, has also been admitted with a fractured tibia after tripping over a coffee table. She lives on her own and rarely goes out, spending much of her day reading and watching television. Why is Ada more likely to have osteoporosis than Sam?

Ada is female, post-menopausal, elderly and takes little weight-bearing exercise. Following repair of the fracture, Ada will need some help to identify how she can improve her mobility and reduce her risk of breaking her bones in the future. Sam is male, young, takes regular exercise and sustained a fracture as a result of major trauma.

All living organisms require energy to do work. The body is amazingly efficient and always tries to conserve energy and materials, avoiding wastefulness. ▣

Homeostasis

To maintain the utmost efficiency, the body needs an optimal working environment. To achieve this the body has numerous mechanisms, most of which require energy, that work to maintain a relatively constant internal environment. This internal constancy is called **homeostasis**. You can think of it as maintaining the right balance of conditions in your body.

TIME TO TRY

Turn back to the tables in Chapter 3 to determine the literal meaning of the word *homeostasis*. _____

Homeostasis literally means 'to stay the same'. Most work done in your body is the result of chemical reactions. These reactions occur most efficiently if there is a relatively constant temperature, the right amount of water and the right amount of chemicals, for example. We have an acceptable normal range of values for each of these. We know normal body temperature is around 37°C. If our body cools too much, chemical reactions will occur more slowly or not at all. If our temperature gets too hot, chemical reactions speed up and some chemicals may be destroyed.

PICTURE THIS

 Imagine you have some prized plants growing in your garden. Before leaving for two weeks during a brutal summer drought, you set a timer on your garden hose that will let it run at just the right flow for two hours each day while you are gone. You return home to find that your once-gorgeous flowers are now dried brown sticks! Some investigation reveals that although your water did run for two hours each day, the flow was reduced because a field-mouse had eaten a hole in your special hose!

1. Why did your plants die? _____

_____

2. In the body, water is a critical nutrient that also carries other nutrients and waste products to and from your cells. What would happen to this transport if you had too little water in your body? _____

3. How might that affect your cells? _____

_____

4. How would overwatering affect your plants? _____

_____

5. How might overwatering affect your cells? _____

_____

The human body is composed mostly of water. It is the main component of your cells and of your blood. Water balance is absolutely critical in the human body. If you have too little water, nutrients cannot be adequately transported to your cells and wastes can accumulate to toxic levels. Your cells will work less efficiently and may die. In plants, too much water dilutes the nutrients the plant cells need and causes them to swell and perhaps die. In the human body, too much water has the same effect, and when brain cells swell, death can occur. This condition is called water intoxication.

Let's consider an example that is not concerned with the body. Think about the central heating system in your home. The thermostat is set at a comfortable level, about 20°C. If the air temperature falls below the set point, the thermostat activates the boiler to produce more heat until the air is back to 20°C, which makes the boiler switch off. Notice that there are two possible problems – the temperature can go too high or too low. Both problems have a solution, but because the problems are opposites, so are the solutions – one raises the temperature but the other lowers it. Both solutions stop once the problem is corrected – they are self-limiting. This kind of control is called **negative feedback**, and it is the most common control mechanism used in physiology and the main way your body maintains homeostasis.

WHY SHOULD I CARE?

Homeostasis and negative feedback are major recurring themes in anatomy and physiology. The normal state of the body provides optimal health. Any significant fluctuation away from the normal range can quickly impair normal function and perhaps become critical. These fluctuations form the basis of disease diagnosis. Negative feedback is the control system by which our bodies correct these errors and restore health, so many of the processes we will discuss in the nervous and endocrine systems will deal directly with homeostatic control through negative feedback.

REALITY CHECK

Let's see what you know about your own body's heating and cooling systems.

If you are overheated, how does your body respond to try to cool you down? _____

If you are too cold, how does your body respond to try to warm you up? _____

If you are too hot, you sweat more and the blood vessels in your skin dilate to bring the overheated blood to the surface of the skin where heat can be more easily lost. If you are too cold, you sweat less and blood stays away from the surface of the skin to keep your internal organs warmer. You shiver, which involves rapid contraction of your muscles. That activity generates heat.

APPLYING THE THEORY

Mr Collins has pneumonia and his temperature is 39°C. He is feeling hot and asks if he can have a fan. This makes him feel cooler but you notice that he is starting to shiver. What is the function of shivering?

It raises the temperature, which is just the opposite effect to the one desired. You need to be aware that cooling the skin has little effect on the inside or core temperature. Although it may make the patient feel more comfortable, don't overdo it.

Balance, or homeostasis, is important in most aspects of physiology. Our bodies strive to stay within normal ranges for blood pressure, pulse, respiratory rate, oxygen and carbon dioxide levels, nutrient and waste levels . . . the list goes on and on. Much of what you will learn in physiology involves the mechanisms by which the body maintains homeostasis. Keep this concept in mind and you will be able to predict more easily how the body might respond in different circumstances and why the body does much of what it does. It's all about balance!

Survival

We have talked about many themes that govern how your body works, but the ultimate aspect is survival. I often tell my students that the body tries to keep us alive, often in spite of our own stupidity. Think about that – most of us can think of occasions when we did something that was not very healthy. Perhaps we overindulged in something we knew was not good for us. Maybe we tried some insane (and useless) fad diet. Perhaps we went too long without sleep or got dangerously overheated while working outdoors in the summer. Although we *are* our bodies and minds, most of the brain functions below the conscious level to coordinate our bodies' activities, and to a great extent this is what keeps us alive even when we abuse ourselves.

You read about homeostasis earlier in the chapter and how these mechanisms are involved in keeping the body on an even keel through feedback systems. Following trauma, the body fights to maintain normality. If someone loses half a litre of blood, the body is able to cope temporarily with the loss of fluid volume. Blood vessels constrict to slow down the loss, heart rate increases to keep the circulation going as normal, blood clots close the wound and fluid is drawn from the tissues into the circulation to maintain the volume. Homeostatic mechanisms can only cope for a limited period before the individual suffers ill effects.

LOOK OUT

Monitor vital signs regularly and frequently in patients who have suffered trauma to ensure that homeostatic mechanisms are coping. Changes should be reported immediately to a senior member of the team so that interventions may be implemented to stabilise the patient if they are starting to lose the battle.

TIME TO TRY

For each of the following situations, think about how your body might respond to try to survive.

1. You are stuck on an island with no food to eat. How will your body get the energy it needs to do work? _____

   _____

2. You've been working outside on a hot day and you're sweating heavily. You are dehydrated. How will your body get your water balance back up? _____

   _____

3. You have a badly infected finger. How does your body prevent you from dying from a bacterial infection? _____

   _____

Our bodies are fairly resilient and rather forgiving. If you have no food, your body will start using the energy stored in your fat and other body materials. Of course, it can only do that for so long before you starve, but most people can last for extended periods by drawing energy from their stores. When you are dehydrated, your brain triggers thirst so that you will replenish your water supply, and it also reduces sweating and urine output to conserve water. If you have a badly infected finger, your body has defence tactics, mostly conducted by your lymphatic (immune) system, that will fight off the infection.

APPLYING THE THEORY

If you have a sore throat, it is not uncommon for the lymphatic nodes to become enlarged in your neck. The system is trying to prevent the infection from spreading to other parts of your body.

There are limits, of course, to how much damage the body can withstand but, overall, the body has numerous control mechanisms and back-up systems that can undo significant damage, self-inflicted or otherwise. Keep this in mind as you study physiology and you will better understand many of the reasons behind body functions. Homeostasis is maintained to ensure survival. We conserve energy and materials to ensure survival. Our body parts perform optimally, making us phenomenally efficient machines with one goal – survival. It underlies almost everything our bodies do.

Reproduction

Did you notice that I ended the last section by saying survival underlies *almost* everything the body does? It is only through reproduction that we pass on our genes and our species survives. Thus even that goal becomes one of survival – survival of more than just you and me. Many biologists maintain that the ultimate goal of our existence is reproduction. Indeed, the vast majority of differences between male and female form and function are tied to reproduction. This is one area in which the body is not so conservative.

Females release at least one ovum, or egg, each month from the beginning of menstruation until menopause, but few of the ova are actually used. Likewise, each month the uterus builds up materials in preparation for a pregnancy that rarely occurs. Then, the newly formed materials are all flushed away as part of the monthly cycle. The levels of several hormones fluctuate in a complicated ballet each month just in case a pregnancy might happen, even for women who are celibate. Men make and release up to a half billion sperm per ejaculation, and if I know my physiology it only takes one to do the job! Now that does seem wasteful. The sexual act itself is also very energy-demanding.

In general, the more energy and materials the body puts into any job, the more important that work is. For this reason, it is clear why many scientists say that reproduction is our ultimate goal. Please don't feel compelled to rush out and do your biological duty – plenty of people out there are doing this well enough to cover for us! The goal is, after all, survival of the species, so some of us can sit out this dance without endangering its continuation. But while you are studying anatomy and physiology, keep this biological goal in mind.

Ultimately, your body functions are focused on your survival, and through reproduction the focus shifts from survival of you, the organism, to survival of the species. ■

✔ **QUICK CHECK**

1. What is homeostasis? _____

2. What is the most common way to regulate homeostasis?

_____

3. Biologically speaking, what is the goal of our existence?

_____

Answers: 1. Maintenance of a relatively constant internal environment. 2. Negative feedback. 3. Survival for the purpose of reproduction.

I Have to Learn All This? Does Anyone Have a **Tissue**?

Your course will probably begin with an overview of the characteristics of life, followed by a discussion of at least some of the basic rules we have just covered. Then, you will move on to explore each of the levels of the Biological Hierarchy of Organisation introduced at the beginning of this chapter. In the remainder of this chapter, we will superficially survey each of these levels. Our plan is to introduce these topics to arm you with some basic knowledge before you go into depth in the classroom. In Chapters 5 and 6 of this book, we will go into a bit more detail on chemistry and cells. Skipping those levels for now takes us to our starting point for the remainder of this chapter: tissues.

Recall that life begins with cells and that groups of cells, in turn, form tissues with specific functions. Tissues must be examined microscopically, and there are four categories:

■ epithelial tissue (epithelium),

■ connective tissue,

■ muscle tissue, and

■ nervous tissue.

Epithelial tissues are like wrappings. They form all linings and coverings in the body. Name any part of the body that has a free surface and I will bet you it is covered by epithelium. The outer part of your skin, called the *epidermis*, is epithelial tissue. The outer surface of your heart, also called the *epicardium*, is epithelium. The linings of all of your hollow organs and your blood vessels are epithelia, although these linings often take on a special name – **endothelium**. *Epi-* is a prefix meaning upon, but *endo-* means inside, and since the epithelium that lines a structure is inside the structure, it is called *endo*thelium.

Connective tissue comes in many types (**Figure 4.5**) and one of its overall jobs, obviously, is to connect. One type of connective tissue, called loose connective tissue or areolar tissue, is basically the body's packing material – it fills spaces between structures and is often specialised into a type of connective tissue called adipose, or fat. Dense fibrous connective tissue is very strong, and forms the tendons that anchor your muscles to your bones and the ligaments that anchor your bones to each other. Cartilage is a shock absorber in many joints, where your bones meet. Bone itself is a connective tissue, as is blood,

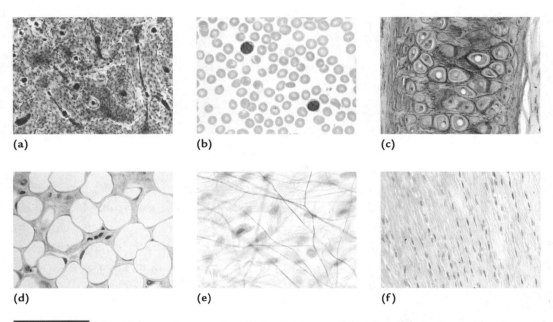

(a)   (b)   (c)

(d)   (e)   (f)

FIGURE 4.5 Some types of connective tissue. **(a)** bone; **(b)** blood; **(c)** cartilage; **(d)** adipose; **(e)** areolar; **(f)** dense fibrous.

Source: (d) from Carolina Biological, Visuals Unlimited. (a), (b), (c), (e), (f) from Pearson Benjamin Cummings.

which is perhaps the body's ultimate connector. There are more types of connective tissue than of any other type, and this diversity allows for a wide array of functions.

APPLYING THE THEORY

Cartilage pads between the vertebrae prevent shocks from walking and running jarring up to the head when you move. They are made of a tough outer layer with a softer gel centre. If too much pressure is placed on one edge of the pad, as happens when you bend forward, it becomes weakened, allowing the gel to bulge out. The bulge may press on a nerve causing pain and muscle spasm. This condition is commonly called a slipped disc and is most likely to happen in the lumbar region leading to pain radiating down the leg on the affected side. This condition is often called sciatica.

This is the reason for so much emphasis on correct moving and handling techniques, both in the health service and industry, so that workers know how to protect themselves from injury.

TIME TO TRY

In what ways are bone and blood both connective tissues – what do they connect? _____

_____

Answer: Bone forms the skeleton upon which the whole body is built, and bones connect to each other and connect your extremities to your trunk, for example. Blood travels through the entire body and connects virtually every cell.

Muscle tissue comes in three types. The one with which you are most familiar is **skeletal muscle**. It provides body movement by attaching to two bones. When a muscle contracts, it moves the bones to which it is connected, thus moving your body parts or your whole body. This is the only type of muscle you can consciously control, so it is called voluntary. **Smooth muscle** is involuntary and is located in the walls of most of your internal organs and your blood vessels. This muscle moves materials through these structures automatically, without you having the mental burden of trying to remember to do it. The most specialised muscle is **cardiac muscle**, which is located exclusively in the heart, and is responsible for the contractions that continuously pump blood throughout your body.

Finally, **nervous tissue** is found in your brain, spinal cord and nerves. This tissue sends signals to your organs and controls their actions. For example, it tells your muscles to contract, causes your glands to release their secretions, and allows you to think.

Can Anyone Here Play **The Organ**?

Now we are moving into areas with which you should be more familiar. Some of the organs you are likely aware of are the heart, lungs, brain, stomach, spleen, kidneys and bladder. Each of these contains more than one type of tissue. Muscles, for example, are anchored to bones by tendons made of connective tissue and are covered by epithelial tissue.

Consider the heart (**Figure 4.6**). Both the outer covering and the innermost lining, as mentioned earlier, are epithelial tissues that

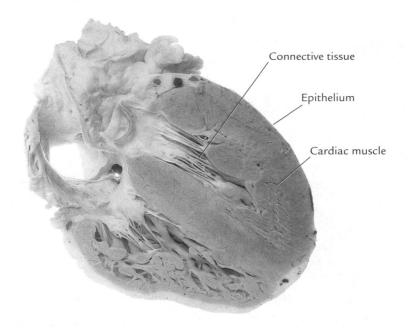

Connective tissue

Epithelium

Cardiac muscle

 FIGURE 4.6   An organ, the heart contains multiple tissues.

provide protection as well as a smooth surface over which your blood can easily flow. The main bulk of the heart wall is cardiac muscle, which contracts. The heart also contains connective tissue, such as the valves that control the direction of the blood flow. Each of these tissues is needed for the heart to move your blood.

✔ **QUICK CHECK**

How can the words *muscle* or *bone* stand for both a tissue and an organ? _____

_____

_____

Answer: Muscle tissue refers to any of three types of tissue that can contract and produce movement. A skeletal muscle is an organ, and it is composed mostly of muscle tissue, but it also has connective tissue (the tendon) and epithelium that covers it. Bone is one type of connective tissue, but bones are organs that contain other tissues as well, specifically other connective tissues and epithelium.

In Sickness and in Health: **The Organ Systems**

As you move through your anatomy class, you will do in-depth explorations of the individual organ systems. Our purpose here is to do a quick summary of each of them (**Figure 4.7**) to give you some background, to get you thinking about how these systems are integrated with each other, and to show ways in which the basic biological rules we previously discussed apply.

The Integumentary System

Your **integumentary system** includes your skin, hair, nails and sweat glands. This system forms your body armour, providing a tough barrier that separates your inside world from the outside. This separation makes it easier to maintain homeostasis. The skin provides protection from potentially damaging substances such as bacteria or harsh chemicals. It helps regulate body temperature. It is also loaded with sensory neurons that gather information about what is going on outside of you, alerting you if there is a potential threat.

The Skeletal System

Your **skeletal system** is primarily composed of your bones, but it also includes the ligaments that hold them together, as well as cartilage and bone marrow. This system is the scaffold upon and around which your body is built. Your muscles attach to it to produce body movement. It also protects some of your most important organs, such as your brain, heart and lungs, thus improving your odds of survival. Minerals, such as calcium, are stored in your bones. If the amount of calcium in your blood drops too low, calcium is released from the bones to maintain homeostasis. And your blood cells are made in your bone marrow, so this system is critical to your cardiovascular system.

APPLYING THE THEORY

Although the skeleton provides protection to the delicate organs, severe knocks can cause fractures, causing damage to the underlying structures. A fractured rib can puncture a lung; a fractured skull can damage the brain. Even if there is no skull fracture, a blow to the head can cause the brain to be smashed against the inside of the hard skull, causing bleeding or bruising resulting in concussion or a more serious head injury. Repeated minor injuries to the head, as sustained during boxing or bungee jumping, can result in brain damage.

LOOK OUT

Carefully observe patients with a history of any type of head trauma for signs of visual disturbance, confusion, drowsiness and nausea, even if there is no wound to be seen. They can deteriorate rapidly, so any change must be reported to a senior member of the team immediately.

REALITY CHECK

List any bones that you know by name. _____

_____

_____

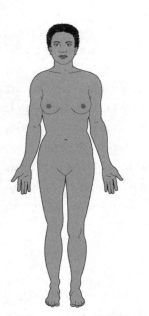

Integumentary system

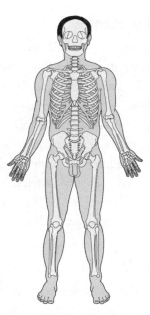

Skeletal system

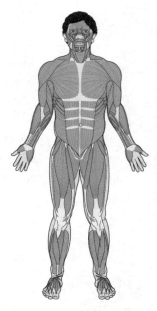

Muscular system

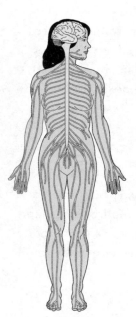

Nervous system

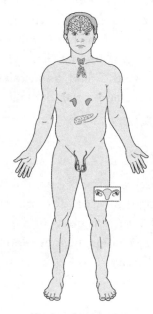

Endocrine system

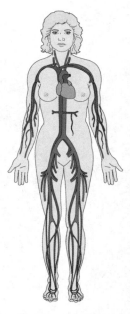

Cardiovascular system

 FIGURE 4.7 The human organ systems.

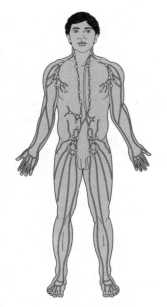

Lymphatic system

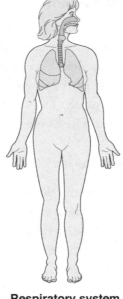

Respiratory system

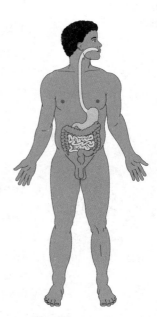

Digestive system

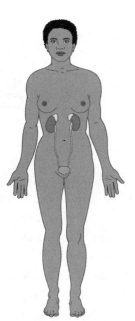

Urinary system

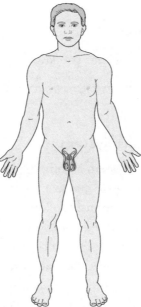

Male reproductive system

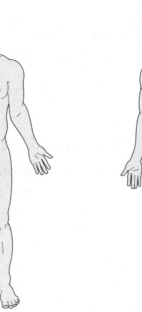

Female reproductive system

The Muscular System

Although there are three types of muscle *tissue*, when we talk about the **muscular system** we are usually referring to the skeletal muscles – the *organs* – and the tendons that anchor them to your bones. Your muscular system, obviously, can move your whole body and also parts of it. Your muscles also protect other structures, and they produce heat, such as when you shiver, which helps maintain your body temperature.

APPLYING THE THEORY

Muscles attached to the skeleton allow movement, but tendons do not have the ability to stretch as muscles can. When the attachment is pulled too rapidly by the contracting muscle, the tendon can tear and the muscle can no longer move the joint. One of the common tears, which often occurs during strenuous exercise, is that of the Achilles' tendon that attaches the calf muscle to the heel.

REALITY CHECK

List any muscles that you know by name. _____

_____

The Nervous System

Your **nervous system** is composed of your brain, your spinal cord and the nerves that go to and from them. This is one of your body's control systems. It continuously collects data in the form of sensory information about everything going on inside you and around you. It decides if any of that incoming information requires a response and, if so, directs that response, perhaps causing muscles to contract to move you away from potential harm, or maybe causing glands to secrete chemicals that digest your food. The nervous system also controls our special senses of taste, touch, smell, vision, hearing and balance. This system is involved in almost every body function and has the major role of coordinating all of our organs and organ systems, keeping them functioning properly and working together to keep us alive.

The Endocrine System

Your **endocrine system** includes your pituitary, thyroid and adrenal glands, your pancreas and your gonads – ovaries or testes – as well as other organs. This system secretes hormones, which are chemical signals that control many aspects of your physiology. It is involved in normal body growth and development, and in maintaining homeostasis, for example, of water and minerals, glucose and blood pressure, to name just a few. Hormones coordinate the functions of your various organs and regulate your body's metabolism, which is the rate at which you use energy. They also control your reproductive activities and cycles. Your endocrine system is intimately linked to your nervous system; in fact, the two systems together reign over most body functions and help you survive.

TIME TO TRY

For each of the hormones listed, write anything you know about what they do in your body.

1. Insulin _____

_____

2. Testosterone _____

_____

3. Adrenaline/epinephrine _____

_____

4. A diuretic causes the body to lose water, usually by increasing urination. What do you think the hormone called *antidiuretic hormone* does? _____

_____

Answers: 1. Insulin regulates blood glucose (sugar) by decreasing it, moving glucose into the cells to be stored for later use. 2. Testosterone is a sex hormone found in highest concentrations in males, causing many sex characteristics associated with maleness, and also contributing to the sex drive in both genders. 3. Adrenaline or epinephrine gives you that adrenaline rush – increased pulse, respiration, blood pressure, and makes you keenly alert and prepared for action in times of crisis. 4. Antidiuretic hormone causes the body to retain water – it has the opposite effect of a diuretic.

Think about how you feel and how someone looks when frightened. You may feel your heart racing, hairs standing up on the back of your neck, muscles become tense, your mouth becomes dry and you have sweaty palms. Your friend may have large pupils and look very pale. These effects are produced by epinephrine (adrenaline) to prepare the body to fight or run away from the situation.

These signs (things that can be measured or observed) and symptoms (things that an individual says that they feel) are also present in patients suffering from shock.

The Cardiovascular System

As mentioned earlier, the **cardiovascular system** includes your heart, blood vessels, and the blood within them. This system circulates your blood, which delivers nutrients, hormones, water and oxygen, to the cells and carrying the waste elsewhere in the body so it can be recycled or discarded. These functions are essential for homeostasis, growth and development, and reproduction. This system also helps regulate your body temperature and provides a common link between all cells. We all know what happens to the organism if the heart stops.

REALITY CHECK

List any blood vessels that you know by name. _____

_____

Are these arteries or veins?

You need to remember that arteries carry blood from the heart to the organs and tissues, while veins return blood to the heart.

The Lymphatic System

Your **lymphatic system** contains your spleen, thymus, tonsils, lymphatic vessels and lymph nodes. This system has two major jobs. It is an alternative return route that picks up extra fluid and materials from your tissues and returns them to your blood, helping to maintain your blood volume and blood pressure. The function which you are most likely to be aware of is your lymphatic system's role in keeping you healthy. This system is also referred to as your *immune system*, and it helps fight infection and disease.

APPLYING THE THEORY

The lymphatic system carries out its drainage work quietly and unobserved until something goes wrong. If the lymphatic vessels are damaged, fluid accumulates in the tissues producing a special type of swelling called lymphoedema. It most commonly occurs in the arm following extensive breast surgery.

The Respiratory System

Your **respiratory system** includes your nose, larynx (voice box), trachea (windpipe), bronchi and lungs. This system is designed to bring air into and out of the lungs so that the blood can drop off the carbon dioxide it picked up from the cells and pick up a fresh supply of oxygen. Blood from your lungs returns to your heart and shoots out through your body so that the oxygen it just gained can be used by your cells to harness energy from your food, allowing your cells to do their work. This cellular process generates carbon dioxide that is picked up by the blood and carried back to the heart to be routed back to the lungs to exchange gases once again – carbon dioxide out, oxygen in. From this, you can see that your respiratory system is directly tied to your cell's energy use. In addition, air that passes across your vocal cords causes them to vibrate, producing sound.

APPLYING THE THEORY

Chronic lung disease is one of the most common reasons for patients being admitted to hospital. It affects their ability to have effective gas exchange, and they often have a high level of carbon dioxide and a lower oxygen level than is desirable. The colour of the blood is less red than normal and the patient's skin may have a blue tinge (cyanosis).

LOOK OUT

Blueness (cyanosis) usually indicates a low level of oxygen, but not all cases are treated by giving the patient more oxygen. Oxygen needs to be prescribed just like any other drug.

JUST FOR FUN

Breathe out onto a cold surface or breathe out into the cold air. What is your respiratory system getting rid of as well as excess carbon dioxide?

LOOK OUT

Water is also lost when we breathe out. We can lose between 500 ml and 1 litre per day in this way, depending on the air and body temperature. Breathing out through the mouth is also a way of losing heat. Some patients can lose a large amount of water if they have a rapid respiratory rate and may need fluid replacement via the intravenous route if they cannot drink enough to maintain normal hydration.

The Digestive System

Your **digestive system** includes your mouth, salivary glands, pharynx (throat), oesophagus, stomach, pancreas, liver, small intestine and large intestine. This system brings food into your body and breaks it down into small molecules that enter your blood. These molecules then provide nutrients, building materials for growth and repair and the chemical energy that your cells use to do your body's work. The digestive system also allows the body to get rid of wastes from the liver, as well as whatever you eat that is not absorbed into your body.

The Urinary System

The **urinary system** includes two kidneys, two ureters, the urinary bladder and the urethra. This system filters your blood, getting rid of waste products and keeping what the body needs, thus maintaining homeostasis. It is crucial for maintaining the proper balance of water, minerals, glucose, blood volume and pressure, and pH in your body. Its role in ridding the body of wastes led to its alternative name – the excretory system.

✔ **QUICK CHECK**

The respiratory, digestive and urinary systems all allow the body to get rid of wastes. How do all three systems do this? _____

_____

_____

Answer: The respiratory system gets rid of the waste gas, carbon dioxide and water when you exhale. Through defecation, the digestive system gets rid of wastes from the liver and the leftovers from eating that our bodies don't absorb. The urinary system cleanses our blood of wastes and excess materials through urination.

The Reproductive System

Finally, your **reproductive system** has different organs depending on your gender. Males have testes, epididymes, the vasa deferentia, seminal vesicles, prostate gland, penis and scrotum. The male reproductive system produces sperm to fertilise the female's ovum, making the father's contribution to a baby's genetic make-up.

The female reproductive system includes the ovaries, uterine tubes (oviducts), uterus, vagina, clitoris, labia and mammary glands. A woman's system is designed to produce the ova that can be fertilised to create a new life, with the mother contributing the other half of the baby's genetic make-up. If pregnancy occurs, her uterus becomes the baby's home during its initial nine months of development, and her mammary glands can feed the baby after it is born.

Together, all of your organ systems carry out all the functions needed to keep you – the organism – alive and, usually, healthy. These systems are all coordinated and interrelated, and illness or malfunction in one often leads to imbalance and malfunction in another. The human organism is, at once, a marvel of complexity and a showcase of simplicity, unsurpassed in its precision by any man-made machine. Take great pride in who you are, for you are a chemical and mechanical miracle!

FINAL STRETCH!

Now that you have finished reading this chapter, it is time to stretch your brain a bit and check how much you have learned.

RUNNING WORDS

At the end of each chapter, be sure you have learned the language. Here are the terms introduced in this chapter with which you should be familiar. Write them in a notebook and define them in your own words, then go back through the chapter to check your meaning, correcting as needed. Also try to list examples when appropriate.

Chemistry
Cytology
Histology
Microscopic anatomy
Gross anatomy
Chemical element
Atom
Molecule
Macromolecule
Organelle
Cell
Tissue
Organ
Organ system
Organism
Energy
Metabolism
Cellular respiration
Calorie
Conservation
Homeostasis

Negative feedback
Epithelial tissue
Endothelium
Connective tissue
Muscle tissue
Skeletal muscle
Smooth muscle
Cardiac muscle
Nervous tissue
Integumentary system
Skeletal system
Muscular system
Nervous system
Endocrine system
Cardiovascular system
Lymphatic system
Respiratory system
Digestive system
Urinary system
Reproductive system

WHAT DID YOU LEARN?

Try these exercises from memory first, then go back and check your answers, looking up any items that you want to review. Answers to these questions are at the end of the book.

PART A: ANSWER THE FOLLOWING QUESTIONS

1. List, in order from most complex to simplest, the levels of the Biological Hierarchy of Organisation that are relevant to anatomy and physiology. _____

   _____

   _____

2. Examine your foot. It transmits your body weight to the ground and supports you in an upright position. How does your foot demonstrate that form fits function?

   _____

   _____

3. Where do we get the energy we use for body processes?

   _____

4. How is energy related to your body weight?

   _____

   _____

5. What is homeostasis and why is it important? _____

   _____

   _____

   _____

6. What are the four general types of tissues?

   _____

   _____

PART B: MATCH THE TERMS ON THE LEFT WITH ALL OF THEIR DESCRIPTIONS ON THE RIGHT

1. _____ muscle   (a) A functional part of a cell

2. _____ lymphatic   (b) An organ

3. _____ epithelium   (c) A tissue

4. _____ organelle   (d) An organ system

5. _____ bone   (e) Covers or lines body parts

6. _____ heart   (f) Skeletal system

7. _____ cardiovascular   (g) Controls most body functions

8. _____ nervous   (h) Keeps you healthy

WEB RESOURCES

Here are some additional online resources for you.

■ *Innerbody.com*

http://www.innerbody.com

This interactive website offers animations, tutorials, and descriptions of items from anatomy. You can click on any organ system for an interactive diagram, then click on most structures to get more information.

■ *Health Information*

http://health.nih.gov

This website from the National Institute of Health is loaded with information and includes a section on body systems.

■ *Histology*

http://w3.ouhsc.edu/histology

The histology laboratory has excellent slides of tissue, which can be viewed at different magnifications. Look at the slides which compare the thickness of skin on different parts of the body.

5 Chemistry
The Science of Stuff

When you have completed this chapter, you should be able to:

■ Understand the different states of matter.

■ Describe atomic structure.

■ Read and understand the periodic table of elements.

■ Explain ionic, covalent and hydrogen bonding.

■ Describe polar molecules and their unique characteristics.

■ Discuss basic organic molecules.

■ Understand that the body is made up of elements and compounds that all work together to make us what we are.

YOUR STARTING POINT

Answer the following questions to assess your chemistry knowledge.

1. The most basic unit of a chemical substance is the _____

2. Matter is defined as anything that _____
_____

3. What are the three states of matter? _____
_____

4. What are the three most common subatomic particles? _____
_____

5. Which subatomic particles interact during chemical reactions?
_____

6. What is the molecular formula for water? _____

7. What are three common types of chemical bonds? _____
_____

8. What happens in anabolic reactions? _____
_____

9. What is meant by *organic* molecule? _____
_____

10. Are proteins organic or inorganic? _____

Yes, we are going to tackle some basic chemistry, but relax – it's really not that difficult. Why do you have to learn chemistry? Anatomy is the study of all of the parts and materials in the body – all of the 'stuff' that you're made of – and chemistry is the science that covers all of that stuff. Physiology is the study of how the body works, and all of the work done in the body involves chemical reactions. Chemistry is very much a part of our everyday lives. Some of you may have previously studied chemistry, but others will be new to this subject. In this chapter, we'll explore basic chemistry concepts to give you a head start for your current studies.

Recall from the last chapter our discussion of the Biological Hierarchy of Organisation. It is organised from the simplest level of organisation to the most complex. The first three levels (atoms, molecules and macromolecules) are part of chemistry so, as you see, chemistry forms the very foundation of anatomy and physiology (**Figure 5.1**).

Learning chemistry may seem tough at times because the terminology can be challenging. To see what I mean, just read the ingredient list on almost any food product. *Do we really eat all that stuff?* But don't let the words interfere with your understanding – much of chemistry is quite simple, even though it may not seem so at first. Here's an example. What do you know about a chemical compound called *dihydrogen oxide?* You probably know more than you realise – that is the technical name for what we usually call *water* – H_2O!

APPLYING THE THEORY

A person essentially consists of a collection of atoms. A nurse needs some knowledge of this truly magnificent, highly organised and very efficient collection of atoms to be able to give high quality and comprehensive care. Knowledge of chemistry helps to understand the normal (physiology) and abnormal (pathology) functions of the body.

What's the **Matter?**

Let's start with something you already know about: **matter**. All the 'stuff' of which you are made is matter, and matter is defined as anything that

- has mass (or weight), and

- takes up space.

The terms **mass** and **weight** are often used interchangeably, but there is a difference. Mass refers to the actual physical amount of a substance. Weight takes into account the force of gravity acting on that mass. Consider astronauts. Each has a certain mass – the actual

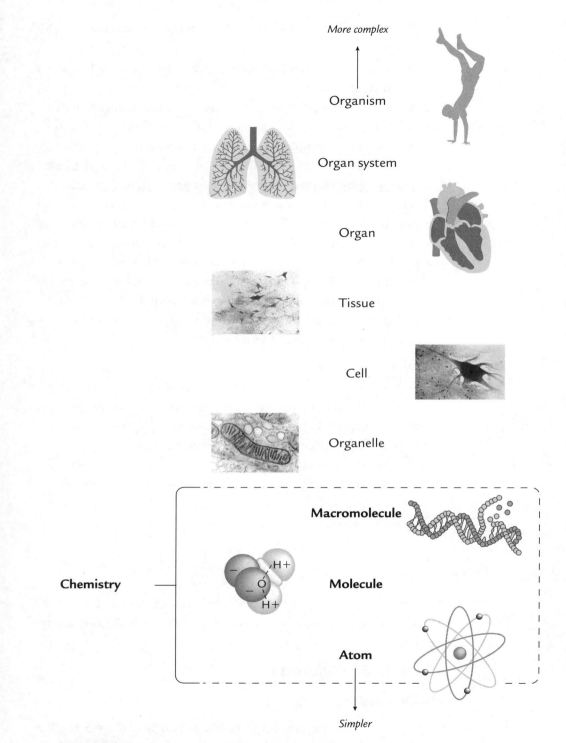

More complex

Organism

Organ system

Organ

Tissue

Cell

Organelle

Macromolecule

Chemistry

Molecule

Atom

Simpler

FIGURE 5.1 **The Biological Hierarchy of Organisation for anatomy and physiology.** The simplest level of organisation is the atom. The first three levels of this hierarchy are part of the discipline of chemistry: the science of matter. All of the other levels of organisation are built upon this chemistry foundation.

amount of material his or her body contains. Each astronaut is weight-less during space travel when there is no gravity, but their individual mass does not change. For the sake of our discussion, matter can be defined using mass or weight, but mass is more precise. The second part of the definition of matter is that it takes up space. The space it occupies is called **volume**.

Matter typically exists in any of three physical states: solid, liquid or gas. Once again let's consider water. What do we call the three states of water?

Solid: _____   Liquid: _____   Gas: _____

I hope you got those! Solid water is ice, liquid water is obviously water, and its gas form is vapour or steam. See – you already know chemistry! Now, how can you change solid water to its gas form? _____

_____

When you add heat, which is a type of energy, ice melts to become a liquid. With enough heat, the liquid eventually boils to become vapour. If you collect the steam and cool it, it will condense back to liquid. If you cool it enough, it will become ice (solid). As you can see, the three forms of matter are interchangeable.

✔ **QUICK CHECK**
What are the three states of matter? _____

_____

Answer: Solid, liquid, gas.

It's **Element**-ary, My Dear Watson!

All matter is composed of chemical **elements**, which are the most basic chemical substances. Over 110 elements are recognised, and around 90 of these occur naturally on Earth. Some elements you prob-ably know are iron, copper, silver, gold, aluminium, carbon, oxygen, nitrogen and hydrogen. Some exist in pure form, such as helium and neon, but most occur combined with other elements.

For the most part, living organisms require only about 20 elements. By weight, 95 per cent of the human body is composed of just four of these:

- carbon,

- hydrogen,

- oxygen, and

- nitrogen.

Each chemical is represented by a symbol, typically the first one or two letters of the element's name. If more than one element name begins with the same letter, the most common of these elements usually gets the single-letter symbol. The symbols for some of the elements, such as the four just listed, are quite logical. Others are less obvious. For example, the symbol for silver is *Ag*, but that is because it comes from the Latin word *argentum*, meaning *silver*. **Table 5.1** lists the names and symbols of some of the elements that are most important for life.

LOOK OUT

Learn the symbols of the main biological elements – you will use them all the time in your work. Blood levels of these elements are maintained within narrow limits so you will need to be able to recognise when they stray from the normal on reports and *remember*: you eat them every day!

TIME TO TRY

Several elements have names that begin with the letter C, so most of them use a two-letter chemical symbol. Try to match each of the following chemical names with their symbols. (*Hint: Recall that one of these is very common and is a major component of all living organisms, including the human body.*)

TABLE 5.1 Some of the important elements in living organisms.

Element Name	Chemical Symbol	Percentage of Body Mass	Role
Oxygen	O	65.0	A major component of both organic and inorganic molecules; as a gas, essential to the oxidation of glucose and other food fuels, during which cellular energy (ATP) is produced.
Carbon	C	18.5	The primary elemental component of all organic molecules, including carbohydrates, lipids, proteins and nucleic acids.
Hydrogen	H	9.5	A component of most organic molecules; in ionic form, influences the pH of body fluids.
Nitrogen	N	3.2	A component of proteins and nucleic acids (genetic material).
Calcium	Ca	1.5	Found as a salt in bones and teeth; in ionic form, required for muscle contraction, neural transmission and blood clotting.
Phosphorus	P	1.0	Present as a salt, in combination with calcium, in bones and teeth; also present in nucleic acids and many proteins; forms part of the high-energy compound ATP.
Potassium	K	0.4	In its ionic form, the major intracellular cation; necessary for the conduction of nerve impulses and for muscle contraction.
Sulphur	S	0.3	A component of proteins (particularly contractile proteins of muscle).
Sodium	Na	0.2	As an ion, the major extracellular cation; important for water balance, conduction of nerve impulses and muscle contraction.
Chlorine	Cl	0.2	In ionic form, a major extracellular anion.
Magnesium	Mg	0.1	Present in bone; also an important cofactor for enzyme activity in a number of metabolic reactions.
Iodine	I	0.1	Needed to make functional thyroid hormones.
Iron	Fe	0.1	A component of the functional hemoglobin molecule (which transports oxygen within red blood cells) and some enzymes.

Your Choices	Names	Symbols
_____	Calcium	Cu
_____	Chromium	C
_____	Cobalt	Ca
_____	Copper (*Latin = cuprum*)	Co
_____	Carbon	Cr

Answers: Calcium = Ca; Chromium = Cr; Cobalt = Co; Copper = Cu; Carbon = C.

APPLYING THE THEORY

Trace elements are required in very small amounts. Many are found as part of enzymes or are required for enzyme activation. Some of the important trace elements in living organisms include copper (Cu), fluorine (Fl), manganese (Mn), selenium (Se) and zinc (Zn). Look at the label on a container of multi-vitamin and mineral one-a-day tablets. Can you see the names of some of these elements in the list of contents?

Chemical Carpentry: **Atomic Structure**

All chemical elements are composed of tiny particles called **atoms**. An atom is the smallest complete unit of an element – one atom of carbon, for example, is the smallest unit, or piece, of carbon that can exist. Two or more atoms can combine together to form larger structures called **molecules**, and simple molecules can join together to form more complex chemical structures called **macromolecules**. These include things like proteins, carbohydrates, DNA and fats – many of the substances we associate with living organisms. But they all begin the same way – made up of atoms.

Atoms vary in size and how they interact with other atoms, but they all share some common characteristics. All are made of smaller units called **subatomic particles** that are arranged in a very precise manner. Although many subatomic particles are now recognised, the main ones of interest are **protons, neutrons** and **electrons**.

The Nucleus

The **nucleus** of an atom is not a structure. Instead, think of the nucleus as the area in the middle of an atom where some of the sub-atomic particles hang out. This can be confusing, because the nucleus of an atom is often referred to as if it is a structure. You should merely think of it as the atom's central region.

An atom's nucleus is where we find two types of relatively large subatomic particles called **protons** and **neutrons.** Protons and neutrons have a similar size and about the same mass. Protons are positively charged particles and may be designated as p^+. Neutrons carry no electrical charge and they are, as their name suggests, neutral. Neutrons may be designated by n^0, indicating they lack any electrical charge, or simply by **n**. All of the protons and neutrons in an atom are located in the nucleus.

Electrons

Orbiting around the nucleus are the other major subatomic particles – the **electrons** – that are in constant motion. Electrons are very small, carry a negative charge, and are often designated as e^-. Because the protons are inside the nucleus, the nucleus always has a positive charge. However, the number of negatively charged electrons orbiting the nucleus always equals the number of protons at the nucleus. Thus the negative charges of the electrons exactly balance the positive charges of the protons. That means that any atom is, overall, neutral.

The number of e^- = the number of p^+. The atom is electrically neutral. ▨

LOOK OUT

Remember A hydrogen atom is the smallest element and always has 1 proton (which carries a positive charge) and 1 electron (which carries a negative charge).

$$p^+ \text{ and } e^- = \text{neutral}$$
$$+1 \text{ and } -1 = 0$$

Electrons are never in the nucleus. Let's start with a simple image to get our bearings. Think of the rings of the planet Saturn: they never touch the planet itself. These rings can represent the paths of the electrons around the nucleus. Unlike Saturn's rings, however, the electrons do not travel in a nice, even, straight line along a single plane. Rather, they buzz about quite rapidly in multiple paths called **orbitals**. For this reason it is more accurate to envision a cloud of electrons that constantly circle the entire nucleus. A picture of an electron cloud does not show actual electrons. Instead, it is more like a map that shows the probability of where the electrons are at any moment.

PICTURE THIS

You arrive home late one evening, well after the sun has set. Your front light is on so you can see to put your key in the lock. You glance overhead and see a large cloud of insects swarming around the light. This is the basic image you should have of the electrons orbiting the nucleus of an atom (**Figure 5.2**). They are constantly in motion around the nucleus, but their individual paths vary.

Now that your skin is crawling from thinking about insects swarming overhead, let's get more specific. The movement of electrons around the nucleus is not as random as the movement of the insects around the light. It is actually a bit complicated, but for our purposes a simplified version will do. Electrons circle around the nucleus at different energy levels, each of which is called a **shell**, and these each have

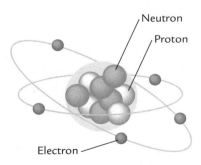

 FIGURE 5.2   Electrons orbiting around the nucleus of an atom of boron.

their own paths around the nucleus. Each shell has its own range of distance from the nucleus, so you can think of each shell as being a specific part of the electron cloud. The first shell is closest to the nucleus. Electrons try to stay as close to the nucleus as they can, but the first shell can only accommodate two electrons. If an atom has more than two electrons, it must have more than one shell. Each additional shell is located a bit further from the nucleus and can hold a maximum number of electrons. Electrons always fill the shells from the nucleus outward. From innermost to outermost the shells can hold 2, 8, 8, 18 electrons and so on. When an electron shell is full, stability of the atom is achieved. To achieve stability, atoms tend to empty or fill their outermost shell by losing, gaining or sharing electrons.

APPLYING THE THEORY

The properties of elements are decided by how many electrons they have in their outermost shell. The fact that electrons go in shells around atoms is the reason for the whole of chemistry. If electrons behaved totally randomly there would be no chemical reactions – nothing would happen. Without shells, atoms would not want to lose, gain or share electrons, so they couldn't form ionic or covalent bonds. But they do have shells, and so the electron arrangement of each atom determines the whole of its chemical behaviour. Electron arrangements explain virtually all of the universe – amazing!!

✔ QUICK CHECK

What are the three subatomic particles, and where is each located in an atom? _____

_____

_____

Answer: Protons and neutrons are always in the nucleus; electrons are always orbiting around the nucleus.

TIME TO TRY

Go to the *Get Ready for A&P for Nursing and Healthcare* website associated with this book, **www.pearsoned.co.uk/getready**

Enter the site and go to the Welcome page. Click on the drop-down box and click on Chapter 5: Chemistry. Then press Go. Read the introductory paragraph and then do the <u>Pre-test</u> quiz.

Can you relate this to what you have just read in your textbook?

Now try the interactive tutorials on <u>*Atoms and Isotopes*</u> and <u>*Build an Atom*</u>.

Atomic Number

Each element has its own **atomic number**. It is, by definition, the number of protons in an individual atom of that element. Each element has a specific number of protons, and all atoms of that element have the same number. For example, hydrogen has an atomic number of 1. If you gave hydrogen another proton, it would no longer be hydrogen – it would now become a different element – helium, with an atomic number of 2. So, in order for atoms to be of the same element, they must all have the same number of protons.

Atomic number = number of protons in an atom. ▨

Now, recall that an atom is electrically neutral overall. This means that the number of positive charges from protons must be counterbalanced with an equal number of negative charges from electrons. In other words, the number of protons in an atom always equals the number of electrons. Because of this, if you know an atom's atomic number, you know not only how many protons it has, but also how many electrons it has – they are the same numbers! Let's try this.

TIME TO TRY

Nitrogen's atomic number is 7.

1. How many protons does an atom of nitrogen have? _____

2. How many electrons does an atom of nitrogen have? _____

If nitrogen's atomic number is 7, that tells you it has seven protons. You know the number of protons must equal the number of electrons, so an atom of nitrogen will also have seven electrons. The seven positive charges from the protons are balanced by the seven negative charges from the electrons, so the atom is electrically neutral. Remember, the first two electrons will go into the first shell and the next five electrons into the second shell around the nucleus.

The number of protons = the number of electrons in an atom. ■

Atomic Weight or Mass

An element's **atomic mass** refers to the total mass of a single atom of that element. Recall our earlier discussion, though, about mass and weight: the terms **atomic mass** and **atomic weight** are used interchangeably. Because you are probably more familiar with weight than mass, we will stick with *atomic weight*. You need to recognise, though, that

1. either of these terms can be used

2. atomic mass is the more precise term, and

3. they have almost the same basic meaning.

For atomic weight we are really talking about how much a single atom weighs. Electrons are so tiny that they weigh almost nothing. Almost all of the atomic weight, then, comes from the combined weights of the larger protons and neutrons. But what do *they* weigh? Obviously atoms are too tiny to be weighed in grams, so imagine trying to weigh a subatomic particle! Conveniently, scientists developed a unit of measurement called the **atomic mass unit (u)**, and 1 u is, to simplify it, about the mass or weight of one proton. Neutrons are almost the same size, so we assume that they have the same weight as a proton. This becomes amazingly simple! To determine an atom's weight, all you do is add its total number of protons and neutrons together. Because each of them weighs 1 u, the atomic weight is the total number times one, and you can simply ignore those tiny little electrons!

Atomic weight = number of protons + number of neutrons in an atom. ■

For all atoms of an element, the number of protons is constant. But the number of neutrons can vary, so atoms of the same element can have different atomic weights. For example, carbon (atomic number 6) has six protons, and it usually has six neutrons, so its atomic weight is usually 12. But some atoms of carbon have seven neutrons, giving an atomic weight of 13, and some have eight neutrons, giving an atomic weight of 14. Adding eight neutrons makes the carbon atom relatively unstable – so it is radioactive. Atoms of the same element that have different atomic weights are called **isotopes**.

APPLYING THE THEORY

Some isotopes are radioactive, meaning they emit certain types of energy. For this reason, some radioactive isotopes are used in medicine. The energy they emit can often be seen with special equipment. For example, the thyroid gland uses iodine to make certain hormones. If a patient might have a thyroid problem, a radioactive isotope of iodine (^{131}I) can be injected into the blood, then the clinician can use an imaging technique to monitor how well the thyroid is working. In another use, cobalt (^{60}Co) can be injected into an area where there is cancer, to irradiate the tumour cells.

Is That An Eye Chart or a **Periodic Table of Elements**?

Look at **Figure 5.3**. Looks a bit scary at first, but that's only because you don't know how to read it. This is the **periodic table of elements**. All chemical elements that are currently known are listed in this table, and more are added as they are discovered. The table is arranged in a specific manner that is quite useful.

From the periodic table, we know that hydrogen's atomic number is 1, meaning it has one proton. Look at **Figure 5.4a** which shows the proton in the centre of the hydrogen atom at the nucleus. This also means it only has one electron, which is shown orbiting the proton, in the first shell. A hydrogen atom is unusual in that it does not have any neutrons. All of the other elements do have neutrons. Helium's atomic

1A																		8A
1 H 1.008	2A											3A	4A	5A	6A	7A		2 He 4.003
3 Li 6.941	4 Be 9.012											5 B 10.81	6 C 12.01	7 N 14.01	8 O 16.00	9 F 19.00		10 Ne 20.18
11 Na 22.99	12 Mg 24.31	3B	4B	5B	6B	7B		8B		1B	2B	13 Al 26.98	14 Si 28.09	15 P 30.97	16 S 32.07	17 Cl 35.45		18 Ar 39.95
19 K 39.10	20 Ca 40.08	21 Sc 44.96	22 Ti 47.87	23 V 50.94	24 Cr 52.00	25 Mn 54.94	26 Fe 55.85	27 Co 58.93	28 Ni 58.69	29 Cu 63.55	30 Zn 65.41	31 Ga 69.72	32 Ge 72.64	33 As 74.92	34 Se 78.96	35 Br 79.90		36 Kr 83.80
37 Rb 85.47	38 Sr 87.62	39 Y 88.91	40 Zr 91.22	41 Nb 92.91	42 Mo 95.94	43 Tc (98)	44 Ru 101.1	45 Rh 102.9	46 Pd 106.4	47 Ag 107.9	48 Cd 112.4	49 In 114.8	50 Sn 118.7	51 Sb 121.8	52 Te 127.6	53 I 126.9		54 Xe 131.3
55 Cs 132.9	56 Ba 137.3	57* La 138.9	72 Hf 178.5	73 Ta 180.9	74 W 183.8	75 Re 186.2	76 Os 190.2	77 Ir 192.2	78 Pt 195.1	79 Au 197.0	80 Hg 200.6	81 Tl 204.4	82 Pb 207.2	83 Bi 209.0	84 Po (209)	85 At (210)		86 Rn (222)
87 Fr (223)	88 Ra (226)	89† Ac (227)	104 Rf (261)	105 Db (262)	106 Sg (266)	107 Bh (264)	108 Hs (269)	109 Mt (268)	110 Ds (271)	111 Rg (272)	112 — (285)	113 — (284)	114 — (289)	115 — (288)				

58 Ce 140.1	59 Pr 140.9	60 Nd 144.2	61 Pm (145)	62 Sm 150.4	63 Eu 152.0	64 Gd 157.3	65 Tb 158.9	66 Dy 162.5	67 Ho 164.9	68 Er 167.3	69 Tm 168.9	70 Yb 173.0	71 Lu 175.0
90 Th 232.0	91 Pa 231.0	92 U 238.0	93 Np (237)	94 Pu (244)	95 Am (243)	96 Cm (247)	97 Bk (247)	98 Cf (251)	99 Es 252	100 Fm 257	101 Md 258	102 No 259	103 Lr 260

FIGURE 5.3 **The periodic table of elements.** An interactive periodic table can be found online at the *Get Ready for A&P for Nursing and Healthcare* website in Chapter 5, Chemistry.

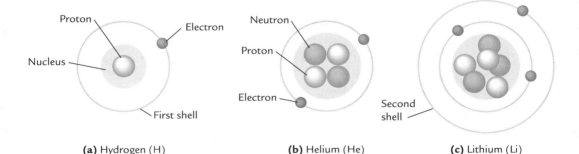

(a) Hydrogen (H)   **(b)** Helium (He)   **(c)** Lithium (Li)

FIGURE 5.4 **Illustrations of atoms. (a)** hydrogen, **(b)** helium, and **(c)** lithium showing the placement of the electrons around the nucleus. The first shell fills first and can hold only two electrons.

number is 2. The helium atom in **Figure 5.4b** has two protons and also two neutrons (so what is its atomic weight? ___). Helium also has two electrons, as shown. Next, look at lithium (**Figure 5.4c**). Its atomic number is 3. You see it has three protons and three neutrons in its nucleus, and three electrons in orbit. The first shell can only hold two electrons, so a second shell is added to hold the third electron.

TIME TO TRY

In the illustrations below, assume that the grey sphere in the middle represents the nucleus and all neutrons and protons in it. Add the shells and electrons for each of the elements below. Boron is done as an example. Don't worry about the positions of the electrons; just draw the right number in the right shell.

Boron	**Oxygen**	**Neon**
Atomic number 5	Atomic number 8	Atomic number 10

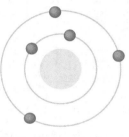

Needs 5 e⁻ total.
2 e⁻ are in the inner shell,
the other 3 in the outer shell.

In your drawings, oxygen should have two electrons in its inner shell and six in its outer shell. Neon should have two electrons in its inner shell and eight in its outer shell, giving it a full outer shell.

LOOK OUT

Remember, if you know an element's atomic number, you know how many protons are in its atoms. Once you know that, you also know how many electrons it has, because the number of protons and electrons is the same.

Now look at **Figure 5.5**. This is the square from the periodic table that represents carbon. You can see that each square of the table tells you an element's chemical symbol, its atomic number, and its atomic weight. What do those three items tell you? _____

_____

_____

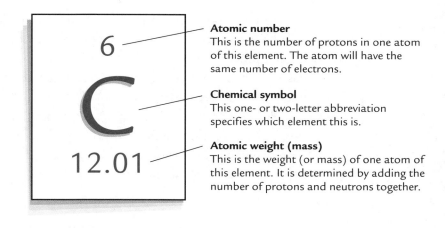

Atomic number
This is the number of protons in one atom of this element. The atom will have the same number of electrons.

Chemical symbol
This one- or two-letter abbreviation specifies which element this is.

Atomic weight (mass)
This is the weight (or mass) of one atom of this element. It is determined by adding the number of protons and neutrons together.

FIGURE 5.5   Information contained in the periodic table.

Now look again at the periodic table. Starting at the top, read it across, left to right, row by row. How is it organised? _____
_____
_____

Next, look at the atomic numbers for the first four elements in the first column – 1 (H), 3 (Li), 11 (Na), and 19 (K). Remember that in any atom the first shell holds only 2 electrons, and each of the next two shells can initially hold up to 8 more. Fill in the missing information:

Element: H   Li   Na   K
Electrons in its outer shell: _____   _____   _____   _____

You should see that all elements in the first column have 1 lone electron in their outer shell. Hydrogen only has 1 electron. Lithium has 2 electrons in the first shell and 1 in the outer shell. Sodium has 2 in the first shell, 8 in the second, and 1 in the outer shell. Potassium has 2, 8, 8 and 1. If you do the same for the second column, you'll find that each element has 2 electrons in its outer shell. And you'll find that all the elements in the last column have 8 electrons in their outermost shells, except for helium, which only has 2 electrons, so their outermost shells are full.

The periodic table is organised by atomic number. The atomic number – the number of protons – increases from the left to the right, and from the top to the bottom. The table is also organised into rows, called **periods**, and columns, called **groups**. Each row, or period, represents a shell of electrons. The first row has one shell, the second row has two shells, and so on. Each column, or group, represents how many electrons are in the outermost shell. This determines the chemical behaviour of the whole group. This organisational approach is very simple to use for the first three rows of the periodic table, but becomes more complicated below that. Those complexities, however, are beyond the scope of our current discussion.

✔ **QUICK CHECK**

What information do you know from the **period** in which an element is found in the periodic table? _____

_____

What information do you know from the **group** in which an element is found in the periodic table? _____

_____

Answers: 1. The period tells you how many shells of electrons there are. 2. The group tells you how many electrons are in the outermost shell.

Bumper Cars and **Chemical Interactions**

Have you ever tried bumper cars? If not, you really should – it's a great way to release tension. Atoms interact with each other rather like bumper cars (**Figure 5.6**). The first part of a bumper car that makes contact with another car is the outer rubber bumper. When two atoms come together, the first parts to make contact are always the electrons in the outer shells. The protons and neutrons are safely tucked away in the middle of the atom at the nucleus. So, the electrons in the outermost shell act as the 'bumper' and determine how atoms interact with each other.

Remember these two points:

1. The number of protons in all atoms of a particular element is constant.

2. The number of protons in an atom = the number of electrons in that atom.

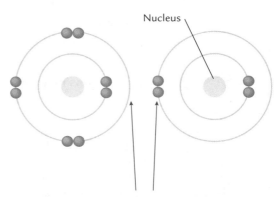

Nucleus

In bumper cars, the outer rubber bumpers of the cars make contact first. The riders inside the car never make contact with each other.

In atoms, the outer shell electrons make contact first. The protons and neutrons inside the nucleus never make contact with each other.

FIGURE 5.6   **Chemicals interact somewhat like bumper cars.** The electrons in the outer shells become the atoms' 'bumpers' and determine the chemical's reactivity.

From this, we see that all atoms of a particular element have the same number of electrons. Because the electrons determine their chemical activity, all atoms of a particular element will react the same way.

TIME TO TRY

Consider carbon in the periodic table.

What is its atomic number? _____

How many protons does it have? _____

How many electrons? _____

Once you see that carbon's atomic number is 6, you know that it has 6 protons, and so it also has 6 electrons. Now draw the electrons for carbon around its nucleus.

Carbon has 2 electrons in the first shell then 4 electrons in its second (outer) shell, and these are the electrons that will interact with other electrons.

The electrons in an atom's outermost shell determine its chemical reactivity. ■

The Union: **Chemical Bonding**

As mentioned earlier, two or more atoms can join together through **chemical bonding** to form a molecule. Recall that the atom's electrons are arranged around the nucleus in one or more shells. The outermost of these subshells is called the **valency shell**. It can contain at most 2 electrons for helium, or 8 electrons for all other elements. If this outermost shell contains the maximum number of electrons, or is full, the atom is amazingly stable. It is said to be chemically **inert** – it will not easily react with other atoms. It is a 'happy' atom. All of the elements in the last column of the periodic table are inert.

On the other hand, atoms of elements in all of the other columns lack a full outermost shell. That means they are unstable and want to become stable. If it helps you remember this, think about life. When we are ful**filled**, we feel happy. If we are not happy, perhaps we feel something is missing from our lives, or maybe we feel we have lots of good to give and nobody to give it to. Now don't you feel sorry for those unfulfilled atoms?

An atom that does not have a full outermost shell of electrons is not stable, and it will react with other atoms to try to fill its outer shell and become stable. Unstable atoms can gain, lose or share electrons with other unstable atoms until they become stable. That's how atoms interact.

✔ **QUICK CHECK**

Under what circumstances is an atom stable? _____

_____

_____

Answer: An atom is stable when its outermost, or valency, shell is full, meaning it has 2 electrons for helium or 8 electrons for neon and argon, for example.

Ionic Bonding

When atoms become stable by losing and gaining electrons, electrons actually leave one atom's outermost shell and join the outermost shell of another atom. The atoms are now stable, meaning they each have a full outermost shell. However, gaining or losing electrons also changes the atoms in another way. Recall that atoms are normally electrically neutral – they have the same number of protons (+) and electrons (−). Once the electrons move, though, the atoms are no longer neutral because the protons and electrons are no longer in equal number. An atom that gains an electron has one extra negative charge, and an atom that loses an electron is one negative charge less than the number of protons, making it positive.

All atoms that have gained or lost electrons carry an electrical charge and are called **ions**. These are designated with a $^+$ or $^-$ sign. For example, sodium tends to lose an electron and become a sodium ion, **Na^+**. Chlorine tends to gain an electron, becoming a chloride ion, **Cl^-**. Ions of opposite charges attract each other (*'Opposites attract'*). Whenever ions are formed, oppositely charged ions will join to form a molecule that is electrically neutral. When they join, they form a strong **ionic bond** – ions form ionic bonds.

Atoms that gain or lose electrons form ions, and ions of opposite charges form ionic bonds. ▪

APPLYING THE THEORY

Ions are charged particles. Ions in solution are called **electrolytes** because they can conduct an electrical current. Electrolytes are a major constituent of all body fluids. They affect many physiological processes and are essential to the normal function of all cells. Common electrolytes include sodium (Na^+), potassium (K^+), calcium (Ca^{2+}), magnesium (Mg^{2+}), hydrogen (H^+) and chloride (Cl^-) ions. Healthcare professionals must fully understand the role of electrolytes in the body and learn to recognise the signs and symptoms of any imbalances.

TIME TO TRY

Let's see how ionic bonding works, using sodium and chlorine. Look at the periodic table and fill in the following information:

	Sodium (Na)	Chlorine (Cl)
Atomic number:	_____	_____
Number of protons:	_____	_____
Number of electrons:	_____	_____

You know from the periodic table that neither of these elements is stable – they are not in the last column of the Table, so their outermost shells are not full. How can they become stable? _____

Draw the electrons around the nuclei of each of these atoms (the shells are drawn for you):

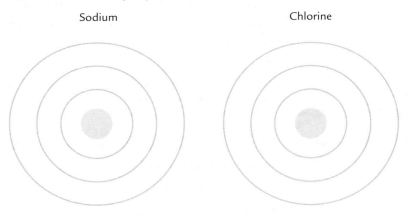

Sodium, with atomic number 11, should have 2 electrons in its inner shell, 8 in the second, and a single electron in its outermost shell. Chlorine, atomic number 17, should have 2 electrons in its inner shell, 8 in its second shell, and 7 in its outermost shell. That is pretty convenient – sodium has one too many and chlorine needs one to fill its outer shell. Sodium will lose its electron to chlorine, producing two ions, Na^+ and Cl^-. Once the ions are formed, their opposite electrical charges will draw them together and they will form a strong ionic bond, creating a molecule of a substance called *sodium chloride*. You know it better as table salt!

Now remember that magic from the periodic table: sodium is in the first column, so it has 1 extra electron it wants to lose. As we saw earlier, all elements in that column have 1 electron in their outermost shell. Chlorine is in the next to last (stable) column. All of the elements in that column need only 1 electron to have a full outermost shell and be stable.

What would you predict about calcium? _____

What about oxygen? _____

Calcium is in the second column, so it has 2 electrons in an outermost shell that wants 8. It is not stable. Oxygen is two columns short of being stable, so it needs 2 more electrons to fill its outermost shell and be stable. Once formed, the ions will combine with an ionic bond to form calcium oxide.

For review, **Figure 5.7** shows how an ionic bond forms between lithium and fluorine. These elements are in the same columns as sodium and chlorine, so the process is the same except these new elements only have two shells of electrons. However, as you now know, in chemical interactions, only the electrons in the outermost shells are important.

✔ **QUICK CHECK**

How does an ionic bond form? _____

_____

_____

Answer: An ionic bond forms when atoms gain or lose electrons, forming oppositely charged ions that are then drawn together by their charges.

Covalent Bonding

Brothers and sisters often argue about possession of some toy or other item. These squabbles usually end with a parental voice from somewhere in the distance yelling for them to . . . *share.*

Apparently some atoms have learned that lesson as well. Let's consider two hydrogen atoms, each of which has a single electron. To form an ionic bond, one hydrogen atom would have to give up its electron and another atom would have to gain it. Which will gain and which

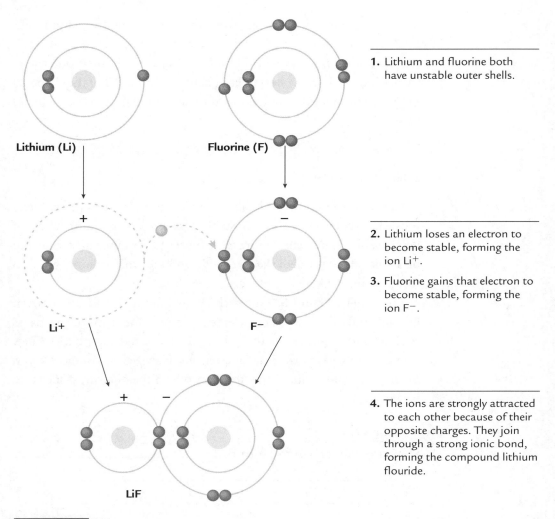

1. Lithium and fluorine both have unstable outer shells.

Lithium (Li)

Fluorine (F)

2. Lithium loses an electron to become stable, forming the ion Li$^+$.

3. Fluorine gains that electron to become stable, forming the ion F$^-$.

Li$^+$

F$^-$

4. The ions are strongly attracted to each other because of their opposite charges. They join through a strong ionic bond, forming the compound lithium flouride.

LiF

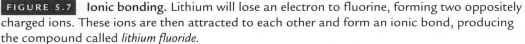

FIGURE 5.7 **Ionic bonding.** Lithium will lose an electron to fluorine, forming two oppositely charged ions. These ions are then attracted to each other and form an ionic bond, producing the compound called *lithium fluoride.*

will lose? Neither; instead, both hydrogen atoms can share their electrons. By combining them, the electrons will orbit around both nuclei together, and both atoms will be stable as long as they stay together. This type of bond is called a **covalent bond**. Recall that the outermost shell is called the valency shell and it contains the electrons that are interacting. The atoms that are sharing electrons in order to have full valency shells to achieve stability are said to be covalent (*co-* as in

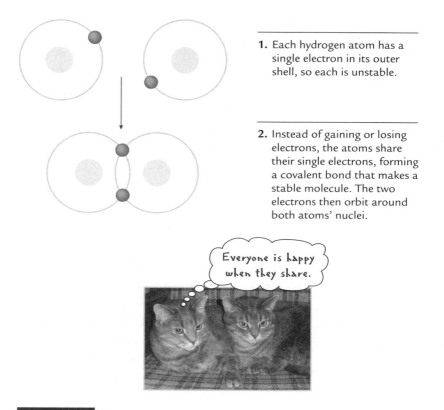

1. Each hydrogen atom has a single electron in its outer shell, so each is unstable.

2. Instead of gaining or losing electrons, the atoms share their single electrons, forming a covalent bond that makes a stable molecule. The two electrons then orbit around both atoms' nuclei.

Everyone is happy when they share.

FIGURE 5.8   Covalent bonding. In covalent bonding, instead of engaging in a tug of war, two atoms share the electrons in their outer shells to become stable. Here a covalent bond between two hydrogen atoms is shown.

together or cooperating). **Figure 5.8** illustrates the formation of a covalent bond. In general, elements that are closer to the right or left side of the periodic table are more likely to form ionic bonds, and those closer to the middle of the table are more likely to form covalent bonds.

✔ **QUICK CHECK**

How does a covalent bond form? _____

_____

Answer: A covalent bond forms when atoms share electrons to attain full outermost shells and become stable.

APPLYING THE THEORY

Carbon is in the middle of the periodic table. It needs to gain or lose 4 electrons to fill its outer shell. This is normally too many for one atom so most organic compounds are formed by covalent bonding (sharing electrons), e.g. methane (CH_4), glucose, and all the carbohydrates, lipids and proteins. This is what makes carbon such a unique element – and perfect for life to be based around its chemistry.

Hydrogen Bonding

Although there are many types of chemical bonds, we will look at just one more. A **hydrogen bond** is a weak bond that can form between the hydrogen atoms in one molecule and some atoms in other molecules. Water is a classic example of this kind of bonding. Look at **Figure 5.9**. A water molecule has two hydrogen atoms and one oxygen atom. A hydrogen atom is tiny, with just 1 electron and 1 proton. An atom of oxygen is much bigger – it has 8 electrons, 8 protons and 8 neutrons. Each hydrogen atom binds to oxygen by sharing its lone electron. Then the electron from hydrogen has to orbit around both nuclei – its own and also that of oxygen. The electrons from both hydrogen atoms will spend more time around the oxygen than around their own nuclei because of the size difference.

PICTURE THIS

Prove this to yourself with Figure 5.9a. Using your finger to represent the path of the shared electron, move your finger at a steady pace to trace the path around one hydrogen atom and move from there to pass around the oxygen atom. Your finger is near the oxygen longer.

In addition, oxygen only shares 2 of its electrons, so it always has 6 around it, even without the shared electrons. So at any given time, there are more electrons around the oxygen than around the hydrogen. As a result, there is a slight imbalance in the electrical charge of the water molecule – the oxygen tends to be a bit more negative than

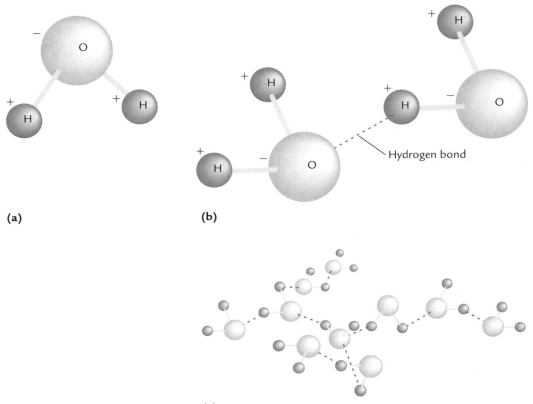

(a)   (b)

(c)

FIGURE 5.9   **Hydrogen bonding. (a)** A water molecule is polarised – the hydrogen atoms tend to be slightly more positive than the oxygen. **(b)** A hydrogen bond forms between a slightly positive hydrogen atom and a slightly negative oxygen atom. **(c)** Hydrogen bonds between water molecules give water many unique properties. Interactive tutorials on hydrogen bonding and nonpolar and polar molecules can be found online at the *Get Ready for A&P for Nursing and Healthcare* website in Chapter 5, Chemistry.

the hydrogens because the electrons spend more time there. This slight charge imbalance is called **polarity**.

Polar molecules can form weak hydrogen bonds, as shown in Figure 5.9b. The slightly more positive hydrogen is attracted to the slightly more negative oxygen in an adjacent water molecule, so they form a weak bond. This is the hydrogen bond. It is a bit like striking up a conversation with a stranger in the checkout queue at the super-market. You happen to be standing close to each other so perhaps you chat briefly, but you are not going to become lifelong friends from that short, superficial exchange. Although they are weak, hydrogen bonds are important. They help give shape to many important molecules, such as proteins, your individual hairs (straight or curly), and the double helix of your DNA.

APPLYING THE THEORY

Water has lots of hydrogen bonds, as shown in Figure 5.9c, giving it many unique characteristics. For example, ice floats because the hydrogen bonds cause the water molecules to spread out more in the solid form than they do in the liquid form. Water also has a high boiling point because of its hydrogen bonds. The fact that life is based on carbon and water is not just by chance but due to their amazing chemical characteristics.

TIME TO TRY

Let's examine the compatibility of polar and nonpolar substances.

1. Take any clear container, preferably one that can be closed. A plastic bag will work. Fill it about a third full with water.

2. Add to that about half as much cooking oil. Try to get them to mix and observe what happens. _____

_____

3. Add a few drops of food colouring to the container and shake it well to mix.

4. Is the food colouring polar or nonpolar (*Hint: into which layer does it settle?*) _____

5. Now find three harmless liquids and try mixing each of them with water. Avoid household cleaners that may be caustic or may react with each other if mixed.

Liquid tested	Polar or nonpolar
_____	_____
_____	_____
_____	_____

You probably found that your liquids, unless they were other oils, were polar, because they mixed with the polar water layer. Water is often used to make **solutions** because so many substances will dissolve in it. In a solution, one substance – the **solute**, dissolves in another substance – the **solvent**. Water is considered a universal solvent, and most chemical reactions in the human body occur in water. Water readily dissolves polar molecules, but it causes nonpolar molecules to bunch together.

Water also has high *adhesion,* meaning it sticks to surfaces very well, and high *cohesion,* meaning its molecules stick to each other. Let's demonstrate that.

TIME TO TRY

You will need two 2p coins, alcohol, water, a dropper (or you can carefully use your finger) and a paper towel.

1. Place the coins on a paper towel and examine them. Estimate how many drops of liquid you can put on a Ip coin before it will spill over. _____ drops

2. Start with the alcohol. Using the dropper or your finger, carefully place drops of the alcohol on the surface of the 2p piece, counting each drop, until it overflows. How many drops of alcohol fit on the coin? _____

3. Now use water and repeat this on the other 2p piece. How many drops of water fit on the coin? _____

You should have been able to pile many more drops of water than alcohol on the 2p coin because the water molecules stick to the 2p coin and to each other much more than the alcohol molecules do, due to surface tension. Because most beverages have a high water content, this property also allows you to fill a glass slightly higher than the rim and observe the surface tension.

✔ **QUICK CHECK**

Explain what is meant by *polar* molecule. _____

Answer: A polar molecule is one in which there is an uneven charge distribution across the molecule, resulting in slightly positive and slightly negative charges.

Molecules and Compounds

As mentioned earlier, when two or more atoms bind together, they form a **molecule**. If the atoms are from the same element, they form a molecule of that element; for example; O_2 is a molecule of oxygen. If the atoms are from different elements, the substance formed is called a **compound**. Water is a compound because it has two elements.

Molecules and compounds are described by a **molecular formula** that includes the letter symbols for the elements and the number of atoms of each element that are present. For example,

H_2O = water

O_2 = oxygen

CO_2 = carbon dioxide

CO = carbon monoxide

LOOK OUT

Remember: the only difference between carbon dioxide (CO_2) and carbon monoxide (CO) is one atom of oxygen. We make carbon dioxide in our body and exhale it with every breath, whereas carbon monoxide is a deadly poison.

The molecular formula gives us limited information. It tells us how many pieces are in a molecule, but not how they are linked together. For that, we can consult the **structural formula**, which is a simplified drawing of how the molecule is built. Lines in a structural formula represent chemical bonds (see Figure 5.9 for water). Now look at **Figure 5.10**. This shows the structural formulae for three sugars: glucose, galactose and fructose. Glucose and galactose are quite similar, so the differences are tinted and boxed. Fructose, also called fruit sugar, has an obviously different appearance.

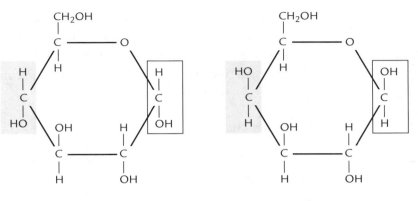

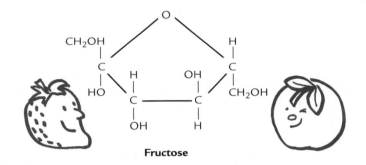

FIGURE 5.10   The structural formulae for glucose, galactose and fructose. Although they all have the same molecular formula, the tinted and boxed areas show the differences between glucose and galactose; fructose is even more obviously different.

TIME TO TRY

Look carefully at Figure 5.10 and fill in the following information.

	Glucose	Galactose	Fructose
Number of carbon atoms:	_____	_____	_____
Number of hydrogen atoms:	_____	_____	_____
Number of oxygen atoms:	_____	_____	_____
Molecular formula:	C__H__O__	C__H__O__	C__H__O__

(*Hint: use the numbers you wrote for each element above.*)

You can see that all three of these sugars have the same molecular formula: $C_6H_{12}O_6$. The structural formula provides more detailed information and is often more useful than the molecular formula, but the molecular formula is the most common method of describing molecules and compounds.

✔ QUICK CHECK

1. What is the basic difference between a molecule and a compound? _____

_____

2. What is the difference between a molecular formula and a structural formula? _____

_____

Answers: 1. A molecule is formed whenever two or more atoms join; if they are from different elements, the substance formed is a compound. 2. A molecular formula tells you how many atoms of each element are in a molecule; the structural formula shows how the atoms are connected.

Love is just a **Chemical Reaction**

All activities that occur within our bodies, including the functioning of our brains and affairs of the heart, involve chemical reactions. When chemicals react with each other, bonds are formed or broken to produce new chemical combinations or to release ions. Energy is stored, energy is released, and energy is used in chemical reactions. Many aspects of physiology involve amazingly complex reactions that are meticulously controlled, whereas others are quite simple.

Chemical reactions are written in the form of a **chemical equation**, but instead of using an equals sign, we use an arrow. The substances to the left of the arrow are the **reactants** – the things that react together. The items to the right of the arrow are the **products** – the end result of the reaction. The arrow means 'produces'. For example, $Na^+ + Cl^- \rightarrow NaCl$ would be read 'sodium ion plus chloride ion produces sodium chloride'. Many biological reactions are **reversible**, meaning they can go in either direction. You can combine the ions we just mentioned to make salt, but salt can also break down to produce the ions. Reversible reactions are often indicated with a special double arrow symbol: $\rightleftharpoons$. Let's examine three basic types of reactions: synthesis, decomposition and exchange.

Synthesis reactions are reactions that build. Another term for these is **anabolic reactions**. With that in mind, think about why athletes sometimes engage in illicit use of *anabolic* steroids – to *build* their muscles and strength. In synthesis reactions, two or more atoms or molecules combine to make a larger molecule. For example, amino acids combine to form proteins and build muscles. Small sugars combine to build large molecules of starch. This is done by forming new chemical bonds between the small molecules (monomers) to hold the pieces together to form large molecules (polymers). Synthesis reactions are especially important in humans for growth and repair processes.

Decomposition reactions are the opposite of synthesis reactions. During decomposition reactions, larger structures are broken down into smaller parts. These are also called **catabolic reactions**. For example, starch is broken down into smaller sugars such as glucose. Salt is broken down to sodium and chloride ions. These reactions are done by breaking chemical bonds. Decomposition reactions are especially important in humans for digesting food and producing energy.

Exchange reactions involve swapping pieces. Two or more molecules split apart and then recombine in a new way; for example, AB and CD separate, then recombine to form AC and BD. Exchange reactions allow the body to receive and store chemicals in one form and then reuse them for multiple purposes.

APPLYING THE THEORY

Physiology is the study of how the body and all of its parts work. All of the work done by the body – collectively referred to as *metabolism* – involves chemical reactions. We eat and breathe to bring in the necessary chemicals, then the body uses those molecules in an amazing array of chemical reactions that allow us to do virtually everything we do. All body processes rely on synthesis, decomposition and exchange reactions. When you begin studying specific chemical processes in the body, it may help to understand them if you think of what the outcome should be – are the reactions building, breaking down, or swapping? If you know the goal, you can more easily understand the reactions.

✔ **QUICK CHECK**

What are the three basic types of chemical reactions? _____

Answer: The three basic chemical reactions are synthesis, decomposition, and exchange.

Look Out – Is it **Organic**?

The term *organic* is used today to describe how food is grown, among other things. In chemistry, though, it has a very precise meaning. By definition, **organic compounds** contain both carbon and hydrogen. They must have these two elements, and they usually have others as well. Most of your body is made of organic molecules. That, Earthling, is what is meant by saying we are a carbon-based life form. The main categories of organic compounds are:

- carbohydrates (sugars and starches),

- proteins,

- lipids (which include fats and steroids), and

- nucleic acids (DNA, which is your genetic material, and RNA, which assists DNA).

Any chemicals that do not contain *both* carbon and hydrogen are not organic, thus they are called **inorganic compounds**. These include carbon dioxide, oxygen, salts and water, which is our most important nutrient.

✔ **QUICK CHECK**

What is the difference between organic and inorganic molecules?

Answer: Organics contain both carbon and hydrogen; inorganics do not.

Our existence requires both organic and inorganic substances. These chemicals are constantly reacting, forming new bonds and breaking old ones, and they account for virtually everything you are and do. Atoms and molecules are the very stuff of life, and the reactions between them are the processes of life. Now you can see that to achieve success in anatomy and physiology, chemistry truly does matter.

TIME TO TRY

Go to *Get Ready for A&P for Nursing and Healthcare* on your computer. Enter the web address **www.pearsoned.co.uk/getready**

Go to Chapter 5: Chemistry. Now do the <u>Post Test</u> Quiz to find out what you have learnt in this chapter.

FINAL STRETCH!

Now that you have finished reading this chapter, it is time to stretch your brain a bit and check how much you have learned.

RUNNING WORDS

At the end of each chapter, be sure you have learned the language. Here are the terms introduced in this chapter with which you should be familiar. Write them in a notebook and define them in your own words, then go back through the chapter to check your meaning, correcting as needed. Also try to list examples when appropriate.

Matter	Electron	Group
Mass	Nucleus	Chemical bonding
Weight	Orbital	Valency shell
Volume	Shell	Inert
Element	Atomic number	Ion
Atom	Atomic weight/mass	Ionic bond
Molecule	Atomic mass unit (u)	Covalent bond
Macromolecule	Isotope	Hydrogen bond
Subatomic particle	Periodic table of	Polar
Proton	elements	Solution
Neutron	Period	Solute

Solvent	Product	Catabolic reaction
Compound	Reversible	Exchange reaction
Molecular formula	Synthesis reaction	Organic compound
Structural formula	Anabolic reaction	Inorganic compound
Chemical equation	Decomposition	
Reactant	reaction	

TIME TO TRY

Now look at the online Glossary in the *Get Ready for A&P for Nursing and Healthcare* website. Make a list of your new vocabulary and check the meanings. You can also make flash cards to help you learn these new terms.

WHAT DID YOU LEARN?

PART A: PROVIDE THE MISSING INFORMATION
CONSULT THE PERIODIC TABLE, FIGURE 5.3

	Potassium	Iodine	Oxygen	Neon
Chemical symbol	_____	_____	_____	_____
Atomic number	_____	_____	_____	_____
Atomic weight	_____	_____	_____	_____
Number of protons	_____	_____	_____	_____
Number of electrons	_____	_____	_____	_____
Number of neutrons	_____	_____	_____	_____
Number of electrons in outermost shell	_____	_____	_____	_____

PART B: ANSWER THE FOLLOWING QUESTIONS

1. Which subatomic particles are always in the nucleus? _____

   _____

2. Which subatomic particles are not included in the atomic weight? _____

   _____

3. Which subatomic particles have a positive charge? _____
_____

4. What are the four main elements that make up most of the
human body? _____
_____

5. In the periodic table, what does the row (period) in which an
element is positioned tell you about that element? _____
_____

6. What does the element's column (group) in the periodic table
tell you? _____
_____

7. Which elements are stable, and why? _____
_____

8. How many electrons can the first shell hold? _____ The
second shell? _____

9. A calcium ion has a +2 electrical charge. How did this ion
form to give it that charge? _____
_____

10. What is the basic difference between an ionic bond and a
covalent bond? _____
_____
_____

11. Which gives you the most information: a molecular formula or
a structural formula? _____
_____

12. When you eat, the food is converted into small, simple
molecules that can be absorbed into your blood. What type of
chemical reactions does that involve? _____
_____

13. If you get a paper cut, what type of chemical reactions will allow your skin to repair itself? _____

_____

14. What elements are always found in organic molecules? _____

_____

15. For each of the following, is the compound organic (O) or inorganic (I)?

$C_6H_{12}O_6$ _____ CO_2 _____

CH_4 _____ CO _____

HCl _____ H_2O _____

WEB RESOURCES

Here are some additional online resources for you.

■ *Headstart in biology*

http://www.headstartinbiology.com

is an introduction to biology for all healthcare students. It was originally designed for students who had just heard that they had been accepted onto a nursing course, and who wanted to use the few weeks before the course began to build a foundation of useful biological concepts. Consequently, the language and references in this resource highlight nursing situations rather than examples from other professions. However, the material and concepts explored are relevant to a range of healthcare professionals and so a broad range of students will be able to use this resource effectively.

Look at this free website – you will need to register.

There are five galleries. Gallery 2 is <u>Atoms and Molecules</u>; Gallery 3 is <u>Chemistry of Life</u>.

There are tutorials on each aspect and a quiz at the end of the gallery.

■ *Get Ready for A&P for Nursing and Healthcare website*

Go to *Get Ready for A&P for Nursing and Healthcare* on your computer. Enter the web address **www.pearsoned.co.uk/getready**

Go to Chapter 5: Chemistry.

■ *BBC Education website – Bitesize Chemistry*

http://www.bbc.co.uk/gcsebitesize/chemistry/

Look at <u>Classifying materials</u> and <u>The periodic table</u>.

■ *Nottingham School of Nursing*

http://www.nottingham.ac.uk/nursing/sonet/rlos/rlolist.php

Try clicking on <u>Atomic Bonding</u> and <u>Elements that make up the human body</u>.

There are many other interactive tutorials here that may interest you.

■ *Wisc Online*

http://www.wisc-online.com/

Sign up for free.

Try <u>Atomic Structure and Ionic Bonding</u> and <u>The Three States of Matter</u>.

6 Cell Biology

Life's Little Factories

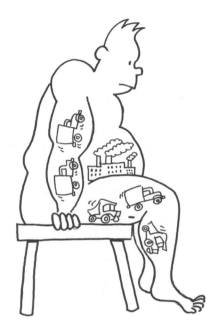

When you have completed this chapter, you should be able to:

■ Explain the Cell Theory.

■ Distinguish between prokaryotic and eukaryotic cells.

■ Describe the structures and functions of the cell membrane and cell organelles.

■ Explain various movement processes that occur in cells.

■ Describe the complete cell cycle and the basics of cell reproduction.

■ Understand that the human body is made up of trillions of cells that all work together and are the lowest level of structure capable of performing all the activities of life.

YOUR STARTING POINT

Answer the following questions to assess your knowledge about cells.

1. Give an example of a prokaryotic cell. _____

2. The liquid located between a cell's membrane and its nucleus is called _____.

3. Another word for *single-celled* is _____.

4. What types of molecules make up the cell membrane? _____
_____

5. What is the function of a ribosome? _____
_____

6. Where in a cell would you find the genes that determine your eye colour? _____

7. What kind of molecules can easily pass through a cell membrane? _____

8. What is *diffusion*? _____
_____

9. What moves during *osmosis*? _____
_____

10. What happens during *mitosis*? _____
_____

Good Things Come in Small Packages: **Cell Theory**

In the last chapter, we explored chemistry and the first few levels of the Biological Hierarchy of Organisation: atoms, molecules and macro-molecules. In this chapter, we continue our climb up the ladder (**Figure 6.1**) as we explore the next two levels: organelles and cells,

Answers: 1. Bacterium. 2. Cytoplasm. 3. Unicellular. 4. Phospholipids, proteins, cholesterol and some carbohydrates. 5. To build proteins. 6. In the DNA in the nucleus. 7. Lipid-soluble molecules. 8. Movement of molecules from an area of higher concentration to an area of lower concentration. 9. Water. 10. The nuclear contents divide.

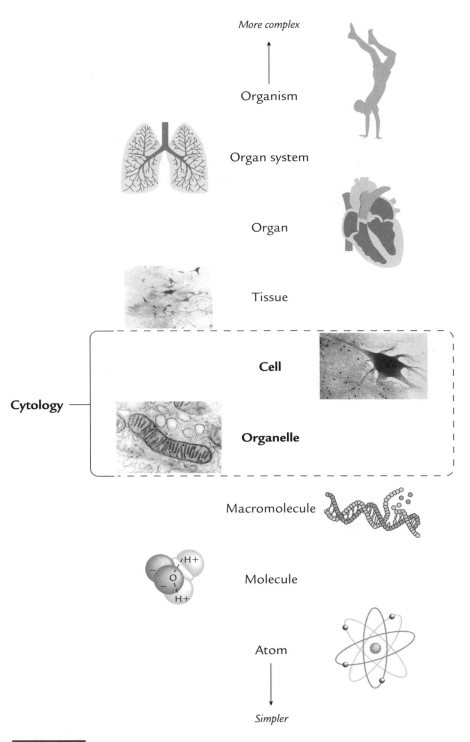

FIGURE 6.1 Moving up the Biological Hierarchy of Organisation, we explore the organelle and cell, both covered in the discipline of cytology.

which are where most of the chemistry occurs in our bodies. This chapter's focus is **cytology**, which is the study of cells (*cyto-* = cell, *-ology* = study of).

In the 1600s, while examining slices of cork with a microscope, English naturalist Robert Hooke noticed that the cork was made of tiny chambers that reminded him of the cells in which monks lived in a monastery. From this observation, he coined the term **cell**, from the Latin word *cella*, meaning 'storeroom' or 'small container'. Since then, vast amounts of cellular research have led to and continue to support a set of conclusions that are collectively referred to as the **Cell Theory**. Let's consider five main principles:

1. All organisms are composed of one or more cells.

2. Cells are the basic structural and functional units of life.

3. All vital functions of an organism occur within cells.

4. All cells come from pre-existing cells.

5. Cells contain hereditary information that regulates cell functions and is passed from generation to generation.

1. **All organisms are composed of one or more cells**. Some organisms are merely a single cell and are called **unicellular** organisms (*uni* = one) and capable of independent life. Organisms made of two or more cells are called **multicellular**. Whether the organism is a bacterium, fungus, plant, or animal (like you), it is made of cells, and the cells of all these organisms show tremendous diversity. Cells come in many different shapes, sizes, and types. In fact, we humans have over 200 different types of cells in our bodies (**Figure 6.2**). Red blood cells, or erythrocytes, are among our tiniest cells, measuring only about 2 μm thick and 7 μm in diameter. (Recall that a micrometre is only 1/1000 of a millimetre.) At the other end of the spectrum, a single neuron (nerve cell) may be up to a metre long! Spermatozoa are minute compared to the ova that they fertilise. Yet, all of the diverse cells found in all organisms share striking similarities and they all originate from a single fertilised ovum.

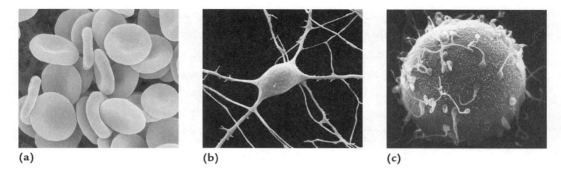

(a)   (b)   (c)

FIGURE 6.2   Human cells come in a variety of types and sizes. **(a)** This scanning electron micrograph clearly shows that red blood cells are biconcave disks. **(b)** A nerve cell (neuron) has numerous cell extensions that can be quite long, as shown in this scanning electron micrograph. **(c)** A human ovum (egg cell) is covered by very tiny sperm cells at fertilisation. *Source:* (a) by Dr. Dennis Kunkel, Visuals Unlimited; (b) © David McCarthy and (c) © David M. Phillips/Photo Researchers Inc.

2. **Cells are the basic structural and functional units of life**. Cells are certainly the basic structural units – all organisms are built from them – but cells are also the functional units of life. Housed within its single cell, a unicellular organism has all of the structures and processes necessary to keep itself alive. Similarly, each cell in a multicellular organism, such as you or me, is an independent living unit capable of maintaining itself. It takes in nutrients, uses them to make the molecules it needs to function, and harnesses energy for doing its work. While going about its daily business, the cell generates and also rids itself of waste. And, in most cases, it reproduces.

3. **All vital functions of an organism occur within cells**. Because each individual cell is alive and carries on all of its own life processes, it is no surprise that these same vital functions carried out at the higher levels of organisation are still done by the cells. As you move up the ladder of complexity, life processes are performed within and around the cells. However, as more cells are added, more organisation is required to meet their demands. As organisms increase in size and complexity, cells become more specialised and join together to share functions, forming tissues and eventually organs. For example, your digestive system brings in adequate nutrients for all of your cells, and your circulatory system ensures that all of your cells are well-serviced and able to

communicate with each other. While your cells perform their highly specialised functions in complex organ systems, your nervous system carefully choreographs them. All the specialisations in structure and organisation keep your individual cells alive and functioning in a highly efficient and synchronised manner, while the cells go about their business of life, maintaining not only themselves, but also you.

4. **All cells come from pre-existing cells**. Cells are capable of reproducing. The normal cell cycle, which we will discuss, ends with the cell dividing to produce two daughter cells. Each of the daughter cells is almost identical to the parent cell and will quickly take on its function. So our body parts are constantly being replenished as old cells die. Similarly, cells divide to replace other cells that may be lost during an injury.

5. **Cells contain hereditary information that regulates cell function and is passed from generation to generation**. Inside each of your cells is the blueprint for life – the DNA that codes the genes in your cells. These genes determine what work your cells will do, and these instructions for life are passed on with reproduction. Not only do the cells divide, but the organism itself can also reproduce through combining special cells. Gametes – ova in females and spermatozoa in males – unite at fertilisation, yielding a new combination of DNA from both the mother and the father. This one cell divides to produce a new offspring in which the whole process of life begins anew at many levels.

✔ QUICK CHECK

What are five principles that make up the Cell Theory?

_____

_____

_____

_____

_____

Answer: All organisms are composed of one or more cells. Cells are the basic structural and functional units of life. All vital functions of an organism occur within cells. All cells come from pre-existing cells. Cells contain the hereditary information needed to regulate cell function and it is passed on to the next generation of cells.

A Cell by Any Other Name . . . **Prokaryotes and Eukaryotes**

As mentioned earlier, cells can be quite diverse. Even within the human body we see tremendous variation in cell form, so imagine how much variety there is if we consider all living organisms! For example, plant cells have rigid cell walls and green chloroplasts, neither of which is found in humans. Despite this tremendous variation, all cells share at least three common characteristics:

1. They are enclosed in an outer **cell membrane**, which separates their internal environment from the external environment.

2. They are filled with **cytoplasm**, which is a mixture of substances in a liquid.

3. They contain deoxyribonucleic acid (**DNA**), which is the cell's genetic material.

Because cells come in many forms, there are many ways to classify them. One classification considers the basic organisation of a cell's internal structure. All cells are either *prokaryotic* or *eukaryotic* (**Figure 6.3**). *Pro-* means before and *karyo-* means '*nucleus*', so the term prokaryote means 'before the nucleus'. Prokaryotic cells lack internal membranes

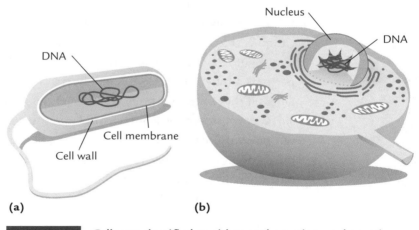

(a)   (b)

FIGURE 6.3   Cells are classified as either prokaryotic or eukaryotic.
(a) Prokaryotes lack internal membranes and have no nucleus, e.g., bacteria. (b) Eukaryotes have both. Examples include all animal and plant cells as well as algae, fungi and protozoa.

so, indeed, they don't have an enclosed nucleus. Instead, their DNA exists mostly in a loop free in the cell cytoplasm, and the area in which it tends to be located (but not enclosed) is referred to as the **nucleoid**. Additional DNA may be present in small loops, called **plasmids**, that can be transferred to other cells. Prokaryotes also lack many other internal cellular structures found in eukaryotes. Although these cells appear primitive, they remain all around us – all bacteria are prokaryotes.

LOOK OUT

All cells have a similar basic structure but it is the differences that make them unique. Both human and bacterial cells contain a lump of genetic material (DNA) surrounded by cytoplasm which is enclosed by a cell membrane. All the other structures make them suitable to function as either a bacterium or a human cell.

APPLYING THE THEORY

Although the human body is composed entirely of eukaryotic cells, prokaryotes are also important. Hundreds of different kinds of bacteria live on or within our bodies. Some bacteria enter our bodies through the foods we eat and take up residence in the colon (large intestines), where they live off our dietary leftovers. While consuming these materials, the bacteria actually produce vitamins that we use. They also, unfortunately, release methane gas in our guts that can cause us to be uncomfortable or, if it escapes, perhaps a bit embarrassed! Some bacteria can cause infection and illness if they enter our body systems. Others occupy space on body surfaces to protect us from harmful micro-organisms as part of our **normal flora**. These are sometimes called **commensals**. Most micro-organisms do not cause disease: the few that do are called **pathogens**. At what temperature do you think pathogens like to live best? Remember, body temperature is 37°C.

The cells of the human body are all eukaryotic cells, or eukaryotes, and share the same fundamental structure – cell membrane/cytoplasm/nucleus. The name comes from Latin: *eu-* means 'true' and these cells have a true nucleus. The nucleus is a membrane-bound structure that houses the DNA of the cell, keeping it localised. The cytoplasm in eukaryotes also has other structures with specialised jobs. Collectively these structures – the functional parts inside the cell – are called **organelles**. Some of these organelles are bound to membranes; others are not. All eukaryotes contain some of the same organelles, but specialised cells often have unique features as well.

All human cells are eukaryotic cells. ▪

APPLYING THE THEORY

Antibiotics such as penicillin work by targeting prokaryotic cells and not affecting eukaryotic cells. In this way, infections caused by bacteria can be destroyed by giving antibiotics with minimal side effects to the human host, as they do not damage eukaryotic cells.

✔ **QUICK CHECK**

In which type of cell would you find a nucleoid, and how is it different from what would be present in the other type of cell?

_____

_____

_____

Answer: A nucleoid is found only in prokaryotic cells and it is where the DNA loop is located in the cytoplasm. Eukaryotic cells, instead, have a nucleus containing DNA that is enclosed by a membrane.

Paper or Plastic? **The Cell Membrane**

Your cells have certain requirements that must be met for them to give their peak performance. For example, they need the right amount of fluid, nutrients, water and oxygen. Recall from Chapter 4 that **homeostasis** means maintaining a relatively constant internal environment (which you can think of as optimal working conditions). Doing so is not always easy.

One feature that helps achieve the goal of homeostasis is the cell membrane, also called the **plasma membrane**. It is simultaneously a container that holds a cell together and a physical partition that separates the inside world of the cell from the outside, making it easier to regulate what enters and leaves the cell. The heading of this section, 'Paper or Plastic', conjures up the image of a shopping bag. That image works for the concept of the cell membrane as a physical separation between the inside and outside of the cell. Clearly, this separation makes it easier to maintain homeostasis inside the cell even if conditions outside the cell vary – keeping your shopping dry even if it is raining.

If the membrane was truly like an actual sack and formed a complete barrier, nutrients could not enter your cells, nor could wastes and products manufactured by your cells leave. Thus, the cell membrane must have unique characteristics that allow some materials to pass through while blocking others.

The cell membrane is made primarily of **phospholipid** molecules. These molecules have phosphorous and other atoms at one end, forming what is called the **phosphate head**. This portion of the molecule is polar, or **hydrophilic**. Recall that term from the discussion in the previous chapter about polar molecules. What does it mean? _____

_____

Attached to one side of the phosphate head are two longer molecules called **fatty acid tails**. These tails are the main reason that the molecule is a lipid, and you should recall from the last chapter that lipids are nonpolar, or **hydrophobic**. What does that mean? _____

_____

To understand how a phospholipid molecule is organised, think about the old-fashioned paper fasteners you probably used in school when you were young or that may even be built into some of your folders now (**Figure 6.4**).

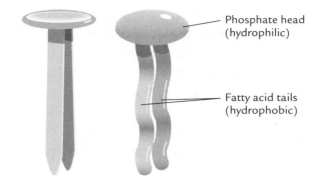

Phosphate head
(hydrophilic)

Fatty acid tails
(hydrophobic)

FIGURE 6.4   **The cell membrane is a phospholipid bilayer.** To understand the organisation of a phospholipid molecule, think of a paper fastener. There is a head with two long parts hanging from it. The head represents the phosphate 'head' of our molecule, which is hydrophilic. The prongs that hang down represent the lipid 'tails', which are hydrophobic.

TIME TO TRY

Let's try to figure out how the phospholipid molecules are arranged in a cell membrane.

1. What happens when a lipid comes into contact with water? Fill a clear container, such as a drinking glass or measuring cup, about 2/3 full of water. Watching closely, pour 1/2 teaspoon of cooking oil into the water. Describe what happens when the oil first enters the water, and where it ends up. _____

   _____

2. Get a slice of bread and cover half of its surface with a thick layer of peanut butter, which has a very high fat (lipid) content. Gently place one drop of water on the bare surface of the bread. What happens to the water? _____

   _____

   Gently place a drop of water on the peanut butter. What happens to the water? _____

   _____

   How do you think these observations relate to the organisation of the phospholipid molecules in a cell membrane? _____

   _____

A cell has an inside and an outside, and both contain water. In humans, the space outside of your cells contains extracellular fluid that is mostly water. The cytoplasm inside your cells also contains large amounts of water. The phosphate heads of the phospholipids interact fine with water, but the opposite ends of the molecules – the fatty acid tails – do not. As you saw when you placed oil in water, lipids in contact with water first form a ball, then settle into a single layer at the surface with one side in contact with the air, not water. In a cell, though, both surfaces are in contact with water. How can phospholipid molecules arrange themselves so the fatty acid tails do not contact water?

Think about that partial peanut butter sandwich you made earlier. You should have seen the water readily enter the bare part of the bread, but ball up on the peanut butter. How could you organise bread and peanut butter in such a way that there are two major surfaces, or sides, and water will not be repelled from either one? _____

I hope you guessed it – you have probably eaten it plenty of times. Consider the basic peanut butter sandwich. Assume you made it with two slices of bread, each smeared with a thick layer of peanut butter, then sandwiched together. You would have bread on both sides of the sandwich and peanut butter in the middle – two layers of it facing each other. This is the basic structure of the **phospholipid bilayer** (*bilayer* = two layers). The hydrophilic phosphate heads (bread) are arranged in two layers so that they face the water in the extracellular fluid and in the cytoplasm. The hydrophobic tails (peanut butter) are sandwiched in the middle, out of contact with the water. This is how the cell membrane is organised (**Figure 6.5**).

Cell membranes don't contain only phospholipids – there are other molecules as well. Cell membranes contain cholesterol molecules, for example, which help stabilise and strengthen the membrane. Some carbohydrate molecules are on the surface and act as labels that allow your cells to recognise other cells, such as when spermatozoa are trying to locate an ovum for fertilisation or when your immune system is destroying foreign invaders to keep you healthy. These are but a few examples.

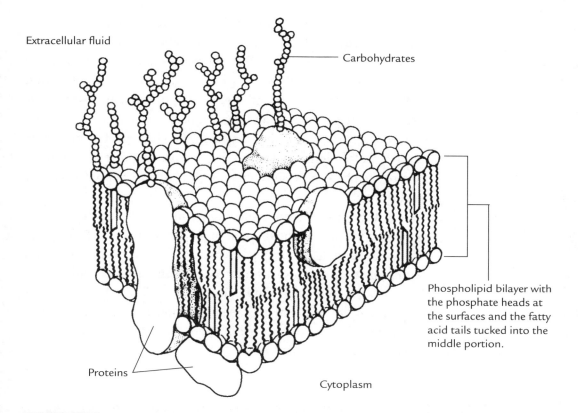

Extracellular fluid

Carbohydrates

Phospholipid bilayer with the phosphate heads at the surfaces and the fatty acid tails tucked into the middle portion.

Proteins

Cytoplasm

FIGURE 6.5   The phospholipid molecules are organised into a bilayer in which other molecules are embedded.

APPLYING THE THEORY

Glycoproteins and glycolipids help cells recognise each other and so act as biological markers. In this way, cells are identified as belonging to your body with a kind of flag, which is known as a self antigen. This is why it is important to try to match blood for transfusions and organs for transplants as closely as possible – to minimise rejection reactions. Self-recognition can go wrong and lead to autoimmune disease. The immune system in the body recognises markers from foreign material, such as bacteria, as not self, and so destroys them.

Cell membranes also contain assorted proteins. Some of these proteins form channels that act like tunnels to allow certain molecules to pass through the membrane. Other proteins might be transporters that pull various molecules through the membrane. The combination of phospholipid molecules and specialised protein channels and carriers determines what can and cannot pass through the cell membrane. Because not everything can pass through, the cell membrane is said to be **selectively permeable** – its chemical composition restricts the movement of some substances, so it is selective about what can pass through.

All of these assorted molecules are positioned throughout the cell membrane in what can be thought of as a sea of phospholipid molecules (see Figure 6.5). Imagine a child's pool filled with water, and floating in it are toy boats, inflatable toys and a few children. The objects are free to move around in the water, and in fact they do. This is the nature of the cell membrane. This model, known as the **fluid mosaic model**, reveals a cell membrane that is very dynamic, moving, changing, and fluid in nature.

The cell membrane is composed of a phospholipid bilayer but also contains other molecules, and it is a very active structure. ◼

JUST FOR FUN

Borrow a child's bottle of bubbles or, if no child is available, make your own loop out of a large paperclip or a length of wire – those twisty ties for rubbish bags work well. Make your own bubble solution by mixing some dishwashing liquid in a bowl or cup of water. Now, blow! Be sure there is plenty of overhead light as you observe the bubbles. Notice how the colours on the wall of a bubble constantly swirl around and change. This is the way the cell membrane is – constantly in motion and changing. See why it is wrong to think of it as just a sack?

What Department Are You With? **Cell Organelles**

Eukaryotic cells contain cytoplasm and a nucleus. The cytoplasm fills the space between the cell membrane and the nucleus. The liquid part of the cytoplasm is called **cytosol**, and it contains many dissolved substances, such as nutrients, enzymes and other materials. It is a rather thick, jelly-like fluid in which numerous cell structures are suspended.

Recall that a typical cell has all the structures and does all the work needed to maintain life. Eukaryotic cells divide the work among different specialised structures called organelles, each with a particular job. To understand this, think about a factory in which some product is manufactured. Inside this factory are many different departments, all involved in some way to make sure the factory is working properly and that the product is made and delivered. In a cell, the departments are the organelles. Although each has its own particular task, all organelles are coordinated and work together. The main structures of a cell are illustrated in **Figure 6.6**.

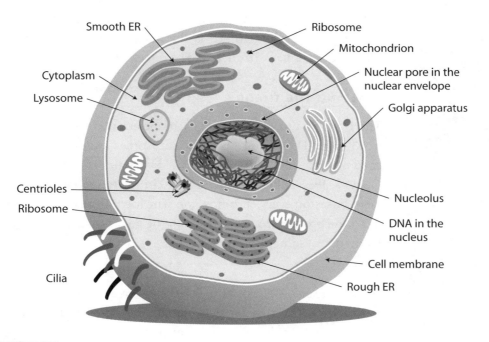

FIGURE 6.6   A composite eukaryotic cell showing the major organelles.

The Nucleus

The **nucleus** is the largest organelle and is present in all cells except red blood cells. It really has one specific task: it houses your genes made of DNA which governs the activity of the cell. Each gene is essentially the instructions for how to make a specific protein. All cells in your body contain two copies of each of your genes – one set from your mother and one from your father. The only exception to this is your sex cells – ova or spermatozoa – which contain only one copy of each of your genes. This copy will combine with a second copy from your partner if fertilisation occurs. Your DNA is organised into thread-like strands, called **chromatin**, that condense into rod-like structures called **chromosomes** when the cell reproduces (more about this later).

Your DNA determines all the proteins that can be made anywhere in your body, but only certain proteins are made by each of your cells. For example, the cells on your elbow don't usually grow hairs, but those on your chin might! The genes are activated and deactivated within the nucleus based on many different factors, and which genes are switched on determines what the cell will do or make. Thus, in your cell 'factory', the office is the nucleus where the boss – your DNA – determines what needs to be done.

APPLYING THE THEORY

There are 46 chromosomes in each nucleus and these are arranged in 23 pairs. There are 22 pairs of autosomes and one pair of sex chromosomes: XY for male and XX for female. The Human Genome Project has deciphered the complete sequence of the DNA bases on every chromosome. The identification and function of all the proteins to be made from the genes is now the ongoing Human Proteome Project.

The nucleus is enclosed in a **nuclear envelope** made of a double membrane similar to the cell membrane. This envelope is pierced periodically by holes called **nuclear pores** that make it more permeable than the cell membrane, allowing larger molecules to pass through. This is important because the instructions – the genes – never leave the nucleus. To get the daily work assignment, molecules must go to the boss's office – the boss does not come to them.

✔ **QUICK CHECK**

What is the main function of the nucleus? _____

_____

Answer: The nucleus houses the DNA.

DNA is your genetic material and it determines what proteins your cells can make. ■

The Ribosome and Protein Synthesis

The DNA in the cell nucleus has the instructions for all proteins that will be made in the cell, but the proteins are not manufactured in the boss's office. A separate cellular department – a tiny organelle called the **ribosome** in the cytoplasm – is responsible for assembling the protein during a process called **protein synthesis**. You can think of a ribosome as being the work bench on which the protein is built, rather similar to an assembly line. The workers who build the proteins there are called **RNA**, which stands for **ribonucleic acid**.

The instructions or work assignments from the boss – the DNA in the nucleus – are carried to the ribosome by a special worker, a molecule called **messenger RNA (mRNA)**. The mRNA carries the work assignment *to* the assembly line. Another worker, a type of RNA called **ribosomal RNA (rRNA)**, is always *at* the assembly line – it is part of the ribosome. This RNA is actually made inside the nucleus at a structure called the **nucleolus**, and it leaves the nucleus through the large nuclear pores.

Once the mRNA delivers the work order from the DNA to the ribosome, the specific proteins are made in the ribosome by linking together small molecules called amino acids. The amino acids are carried to the ribosomal assembly line by other workers called **transfer RNA (tRNA)**, then assembled according to the DNA instructions. So the job of RNA is to build the protein the DNA tells it to make. Messenger RNA carries the instructions from the DNA to the ribosome, and ribosomal and transfer RNA are directly involved in building the protein according to the genetic code of DNA.

The Endoplasmic Reticulum

Some ribosomes are free to float around the cytoplasm making soluble proteins; others are attached to yet another organelle called the **endoplasmic reticulum**, or **ER**. The ER is an extensive network of membranous tubes and channels inside the cell. Ribosomes look like tiny dots, so when they are attached to the ER they give it a rough appearance, thus it is called rough ER. The presence of ribosomes tells you that at least part of rough ER's job is protein synthesis. Smooth ER lacks ribosomes. Instead of making proteins, it is involved in making other materials (like lipids), detoxifying potentially harmful substances, and transporting materials around the cell.

You can think of the ER as being like a system of workers who are always passing through various hallways, moving materials from one work station to another, and throughout the factory. Because the ER connects different parts of the cell, it also provides a communication network within the cell. Materials are brought into this system, moved around, changed, and turned into various products here.

Golgi Apparatus

The **Golgi apparatus** looks like a stack of flattened membranous sacs. This is the processing, packaging and shipping department in our cells. Products that have been made elsewhere in the cell, such as at the ribosomes or in the ER, are sent here to be finished. They are modified and put into their final forms. They are given a molecular 'shipping label' – a chemical tag that determines where they will go. They are packaged in a membrane that pinches off the Golgi apparatus, forming a sac-like structure called a **vesicle**. Finally the products are shipped, some to other parts of the cell, some to the cell membrane, and some out to the great extracellular world beyond.

✔ **QUICK CHECK**

1. Where in a eukaryotic cell would you find ribosomes? _____

   _____

2. What is the functional relationship between the nucleus, nucleolus, ribosome, rough endopalsmic reticulum and Golgi apparatus? _____

   _____

   _____

Answers: 1. Ribosomes may be free-floating in the cytoplasm or attached to endoplasmic reticulum. 2. The nucleus houses the DNA that has instructions for how to build proteins. Proteins are built at the ribosome, part of which is made in the nucleolus inside the nucleus. Many of the ribosomes are located on the rough ER. Proteins made at the ribosomes move into the Golgi apparatus to be processed, packaged and shipped to their final destinations.

The Mitochondrion

If your factory is to do its work 24/7, it needs a good power source. The **mitochondrion** is the powerhouse of the cell. It provides a constant supply of energy to drive the work being done throughout the cell. Cellular energy comes from the foods we eat, where it is stored in the chemical bonds that hold the atoms of the food together. Specialised chemical reactions involving aerobic cellular respiration in the mitochondria harness that energy and store it in a molecule called **ATP** – *a*denosine *tri*phosphate (see why we call it ATP?). All cells, in any organism, use ATP directly for energy. Think of it as the electricity that powers our cells. The more work a cell is doing, the more ATP it needs, and the more mitochondria it will have (e.g., muscle cells).

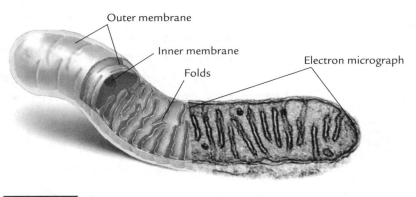

Outer membrane

Inner membrane

Electron micrograph

Folds

FIGURE 6.7   **A mitochondrion.** This illustration includes a drawing (left) blended into an actual electron micrograph.
Source: Pearson Benjamin Cummings (drawn part); © D.W. Fawcett/Photo Researchers Inc. (electron micrograph).

Mitochondria are unique organelles (**Figure 6.7**). They contain their own DNA and can reproduce. These elongated organelles are enclosed by a double membrane, similar to the nuclear membrane. The outer membrane is smooth, but the inner membrane is highly folded. In general, such folding occurs to increase the surface area of the membrane, and increasing the surface area is the same as increasing the work space. It is rather like expanding a department.

TIME TO TRY

To understand how the foldings of the mitochondrial membrane increase the surface area, get a small plastic storage bag. This represents the outside membrane of the mitochondrion. Notice that it is smooth all the way around. If you put another smooth membrane inside of this bag, it would be about the same size. Now, take a larger plastic grocery bag and note how much larger it is. This will represent the inner folded membrane of a mitochondrion. Start folding it. Can you fold it enough to fit inside of the original bag? If you fold it enough, it should all fit inside. Now realise how much more membrane can be used if it is highly folded.

Lysosomes

While the organelles are doing all this hard work in your cellular factories, they need some assistance in keeping their workplace tidy. **Lysosomes** are your cell's cleaners. A lysosome is a small spherical membranous bag containing strong digestive enzymes. Its main job is to break down materials. Some materials are brought into your cell and digested to provide basic building materials for your cell to use in its work. Other materials may be old, worn out cell parts or foreign material (such as bacteria and viruses) that can invade your cell. Lysosomes destroy these items so they neither harm nor clutter the interior of the cell. A lysosome's job is to recycle what it can and get rid of the remaining waste products.

LOOK OUT

Lysosomes contain enzymes that break down toxic components – so there are large numbers in liver cells, reflecting one of the liver's main functions, that of detoxification. Compare this to the concept that large numbers of mitochondria are found in cells that need a lot of energy. *Remember* that the structure of a cell follows the function that is required of it. Would a heart cell contain more lysosomes or more mitochondria? (Remember myocytes need lots of energy – so the answer is mitochondria.)

The Cytoskeleton

Remember that all of this work is going on inside your cells, in a liquid environment. The 'building' needs a frame to hold it up. We discussed the outer partition – the cell membrane – but we need something inside to hold the membrane out so the cell does not collapse on itself. The **cytoskeleton** is composed mostly of tiny tubes (**microtubules**) and protein rods (**microfilaments**). These structures form a type of scaffolding that supports the cell and to which various organelles are attached. Although the name sounds like this structure is made of bone, the cytoskeleton is actually made of proteins. You can think of them as being the struts and beams that hold up the building, or cell. Any breakdown in the links is potentially lethal, e.g., muscular dystrophy.

Cell Movement: Centrioles, Cilia and Flagella

We have discussed how materials can move through a cell, but there are other movements associated with a cell, too. An area of the cell called the **centrosome** ('central body') is composed of paired cylindrical structures made of microtubules. These structures are called **centrioles**. They direct the movement of the chromosomes when a cell reproduces, as we will discuss shortly, but they also form the bases of two other structures involved in cell movement: cilia and flagella.

Cilia look like fringes on a cell. Not all cells have cilia, but those that do have a lot of them. They are extensions of the cell and they are mobile. The movement of cilia is coordinated so that they tend to move in a wave-like manner. Cilia sweep materials over the outer surface of a cell, moving materials past the cell. For example, in the respiratory tract the cilia help clear debris so it doesn't get into and clog the air sacs where oxygen enters our blood.

PICTURE THIS

Imagine you are at the 'big game' and the crowd is tossing a beach ball around. As this is going on, off in the distance you see the crowd start a 'wave', where they stand and wave their arms overhead, then sit, group by group, all around the stadium. The wave approaches you just as the beach ball is heading your way. The wave passes you and so does the beach ball. It was carried away on the wave and you see it now making its way around the stadium, riding the wave. This is how cilia move materials across the surface of the cell – they beat in a coordinated manner, sweeping materials along.

APPLYING THE THEORY

Cigarette smoke paralyses the respiratory cilia for about an hour per cigarette. During that time, they cannot prevent the particulate material that we breathe in from reaching deeper into the lungs. Over time, more of this material clogs the small air sacs and begins to damage them. People who experience this must cough, especially upon arising in the morning, to try to clear the material that has accumulated in their lungs. This is the basis of *smoker's cough*.

In contrast to the cilia, a **flagellum** is a single, long, tail-like extension of the cell. In humans, these are found only on spermatozoan cells. A flagellum whips back and forth to propel the sperm through the male reproductive tract and up into the female's tract in search of an ovum to fertilise.

We have reviewed the organelles of a typical eukaryotic cell, like our human cells. You will learn more detailed descriptions of these structures and their jobs as you proceed in your A&P course. For now, you should have a general understanding. **Table 6.1** summarises the organelles we have discussed and provides a quick review.

✔ **QUICK CHECK**

1. Flagella use tremendous amounts of energy to propel the cell forward. Which organelle would you expect to see in large numbers near the very busy flagellum? _____

2. How do the movements of cilia and flagella differ? _____

_____

_____

Answers: 1. Mitochondria. 2. Cilia are numerous and they beat in a wave-like manner to sweep materials across the cell surface; a flagellum is a single whip-like process that propels the entire cell.

TIME TO TRY

Go to the *Get Ready for A&P for Nursing and Healthcare* website associated with this book, **www.pearsoned.co.uk/getready**

Enter the site and go to the Welcome page. Click on the drop-down box and click on Chapter 6: Cell biology. Then press Go. Read the introductory paragraph and then do the Pre-test quiz.

Can you relate this to what you have just read in your textbook?

Now try the interactive tutorials on Cell Membrane and Cell Organelles – Nucleus.

TABLE 6.1 Summary of cell organelles and structures.

Organelle	Description	Function
Nucleus	Rounded larger membranous sac with pores.	Houses the DNA that directs cellular activities.
Chromatin	Relaxed strands of DNA in the nucleus.	Contains genes that determine what proteins can be made in the cell.
Ribosome	Small non-membranous structure free in the cytoplasm or attached to ER.	Provides the site for protein synthesis.
Nucleolus	Small body located in the nucleus.	Makes part of a ribosome.
Endoplasmic reticulum (ER)	Extensive network of membranous tubes and channels.	Protein synthesis (rough ER); lipid synthesis (smooth ER); detoxification; communication and transport system.
Golgi apparatus	Flattened stack of membranous sacs.	Processing, packaging and shipping of cellular products.
Mitochondrion	Elongated membranous structure with highly folded internal membrane.	Powerhouse; harnesses energy from food molecules and stores it in ATP.
Lysosome	Small membranous sac.	Breakdown of unwanted materials; recycling of molecules.
Cytoskeleton	Meshwork of microtubules and microfilaments.	Provides support and structure to the cell interior; anchors organelles.
Centrioles	Cylindrical structures made of microtubules.	Direct movement of chromosomes during cell reproduction; form part of cilia and flagella.
Cilia	Small, numerous, hair-like processes that beat in a wave.	Sweep materials over the surface of the cell membrane.
Flagellum	A single whip-like tail.	Propels the cell forward.
Vesicle	Small membranous sac.	Contains materials entering or leaving a cell.

Tunnels and Doorways: **Movement Processes**

We have discussed how cilia and flagella produce movement for cells, but these are not the only movement processes occurring. Atoms and molecules are always moving, both inside and outside the cell. Cell products packaged at the Golgi apparatus are shipped out of the cell to travel elsewhere in the body. Nutrients and building blocks are moved in from the outside. Waste products must leave. Let's examine how some of these movements occur.

Brownian Motion

Atoms and molecules constantly move in a random manner called **Brownian motion**. This is a rather non-directional, jiggling movement. Remember those bumper cars we had fun with before? If you are in a car and someone hits you, their car bounces off yours in a new direction, while your car careers off in yet another direction. This is similar to how atoms and molecules move. Because they are constantly in motion, you can imagine how the particles will occasionally bump into each other and ricochet away.

Concentration Gradients and Equilibrium

Although Brownian motion is random, other movement processes are not. Before we discuss these movements, though, we need to understand the concepts of concentration gradients and equilibrium. You can think of concentration as referring to how crowded together molecules are – the more crowded they are, the more concentrated they are. A **concentration gradient** exists whenever there is a difference between the concentrations of the molecule in two areas (**Figure 6.8a**). Perhaps your vehicle has gradient glass in its windscreen – it is tinted with more colour at the top than at the bottom, which means there is a *gradient* in the colour.

If, instead, there is about equal space between all of the molecules, we say they are at **equilibrium (Figure 6.8b)**. Think about standing in a lift. If there are only two of you in the lift, you will probably stand on opposite sides. Add two more people and you will probably stand one in each corner with about equal distance between yourselves. As you add more and more people to the lift, you become more closely crowded

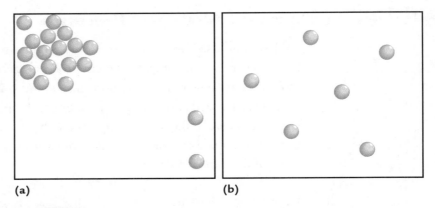

(a)   (b)

FIGURE 6.8   **(a)** A concentration gradient exists when there is a difference in the concentration, or spacing, around molecules in two different areas. **(b)** Equilibrium exists when the molecules are spaced about evenly.

and are more likely to bump into each other. If you do collide, you respond by moving away, always trying to maximise your personal space. That is how atoms and molecules move, but not by conscious decision. They move by basic laws of science. Let's explore how.

Simple Diffusion

One of the most basic types of molecular movement is **simple diffusion**. This is how oxygen enters your cells and carbon dioxide leaves them, for example, and it is critical for sustaining life. Simple diffusion occurs when molecules move from an area of higher concentration to an area of lower concentration. In other words, they move from where they are more crowded to where they have more room. You can think of it simply as the molecules spreading out. Whenever there is a concentration gradient, molecules will spontaneously move *down the concentration gradient* – that means they move from where they are most concentrated, or crowded, to where they are least crowded. You are already familiar with diffusion. Consider, for example, baking brownies. The molecules that produce the lovely smell quickly diffuse through the air in your house so that any visitors know you have a treat to share!

With individual molecules the motion is random but, like people in an elevator, the more molecules that are present in an area, the more likely it is that they will bump into each other and bounce away. There will be more collisions in an area where the molecules are more crowded, sending them skidding off. They will ricochet less as they move into areas where there are fewer molecules to bounce off. In time the molecules, like the people in the lift, will have moved around enough to have almost equal distance between them. This is equilibrium. However, unlike the people in the lift, carefully maintaining their positions, molecules at equilibrium do not stop. They continue moving, but all molecules experience about the same number of collisions and they maintain a fairly even spacing.

The rate at which molecules diffuse varies under different conditions. Molecules diffuse faster when there is a greater concentration gradient between the two areas. Molecules in high concentrations will diffuse faster than those in lower concentrations. Smaller molecules also move faster than larger molecules, and temperature will alter the diffusion rate as well. Diffusion is fast over short distances, which may explain why cells have evolved to be so small.

TIME TO TRY

Let's see how temperature affects the process. Take three clear glass containers of about the same size. Fill one halfway with very hot water, another with water near room temperature, and the third with very cold water. Line them up and wait until the water stops moving. Gently add one drop of food colouring to each container. If you have no food colouring, you could use a tea bag. Now just observe. You should see evidence that the molecules are diffusing – the colour should spread out from where it is most concentrated. Eventually you should also see the equilibrium state – all of the water should be of uniform colour, meaning that the molecules have spread out equally, even though they continue to move.

How did the diffusion rates differ between the three glasses, and why? _____

_____

You should see that increased temperature also increases the diffusion rate. This is because heat makes molecules move faster.

APPLYING THE THEORY

Cell membranes are called **selectively permeable** as they select which molecules pass through them. Factors affecting the rate of diffusion include temperature and pressure/surface area and thickness of the membrane/size of the molecule to transport, its concentration and charge. Many conditions in the lungs are associated with poor diffusion of gases. Emphysema involves an increased size of the alveoli, so decreased surface area of lung, hence less gas exchange. Pulmonary fibrosis concerns the increased thickness of the membrane of the alveoli, leading to a decreased gas exchange, which means there is less oxygen passing into the blood and accumulation of carbon dioxide.

Facilitated Diffusion

Simple diffusion is only one mechanism that allows materials to enter and leave a cell. Only lipid-soluble nonpolar molecules can diffuse directly through the cell membrane. Larger polar molecules, such as glucose (one type of sugar) cannot diffuse through the membrane as easily. Instead, they are moved by a special protein carrier molecule in the cell membrane. The molecules still move from an area of high concentration to one of low concentration, trying to reach equilibrium. The carrier molecule merely helps, or facilitates, the movement of the molecule through the cell membrane. It acts like a special door through which they can pass. This type of diffusion is called **facilitated diffusion**.

The hormone insulin is involved in the transport of glucose and affects the rate at which glucose enters the cell.

Diffusion is a passive process by which molecules spontaneously move from where they are in high concentration to where they are in low concentration. ■

Osmosis

Osmosis is a special type of diffusion that is also critical to our survival. Biologically, it is the diffusion of water through a selectively permeable membrane, such as the cell membrane. We have discussed the lipid nature of the cell membrane, so it seems surprising to discover that water actually passes through it fairly well. This is partly due to the fluid nature of the molecules in the membrane, so before proceeding, we need to recall the basics of solutions.

TIME TO TRY

Let's go into the kitchen and get a clear drinking glass and fill it about 2/3 full with warm water.

Describe what you see. _____

Next, get a spoonful of sugar. Describe what you see. _____

Finally, add the sugar to the water and mix it thoroughly. What do you see? _____

This exercise may have seemed silly, but you just made a solution by combining a solute and a solvent. The sugar is the **solute** – the material that gets dissolved. As in our bodies, water is the **solvent** – the material that dissolves the solute. The end result is a **solution**, which is a homogeneous mixture – it should look uniformly clear throughout, because you cannot see the dissolved solute.

To understand the relationship between the water and solute molecules, recall that matter is anything that has mass and occupies space, so no two molecules can occupy the same space at the same time. Let's think about that sugar solution you just made. If you started by first putting a half-cup of sugar in the container, there would have been less room left to fill with water. The opposite would also be true – less sugar would mean more water. It doesn't matter what the solute molecules are – they each take up their own space (**Figure 6.9**). Let's adopt a highly technical term for the various solutes: *stuff.* When we are

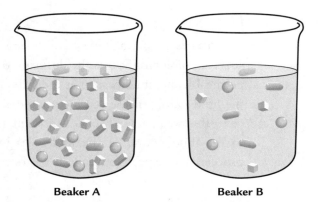

Beaker A   Beaker B

FIGURE 6.9   A solution contains both a solvent, such as water, and solute molecules. In this illustration, the beakers contain the same amount of solution. The geometric shapes in each solution represent various solute molecules. The remaining shaded space represents the water. Clearly, beaker A, which contains more solute molecules, has less space left for water, but beaker B, with fewer solute molecules, contains more water.

talking about osmosis through a cell membrane, we must look at the water and the other stuff for a simple reason – the more stuff you have, the less water there can be because the stuff takes up space that water can no longer occupy.

TIME TO TRY

Prove to yourself that when there are more solute molecules in a given space, there must be less solvent and vice versa. In other words, the more concentrated the solute, the less concentrated the solvent (water). Get a glass measuring jug, then grab a handful of household items – paper clips, coins, marbles and so on. The only rules are that the objects cannot float and they cannot dissolve (most food items are not good for this reason). These objects represent the solute molecules – the stuff – and the water, of course, is the solvent. Drop all of the objects into the jug, then fill it with water to a marked level. Look at the layers of water and objects and note the sizes of each layer. Carefully pour the water into a second container and put it aside. Place the objects on some paper towels, then pour the water back into the jug to measure it. How much water was there in the solution? _____

Repeat this procedure, but this time place only about half of the items in the jug before filling it with water to the same level. Again, pour the water into your second container, remove the objects, then pour the water back into the jug to measure it. How much water did your mock 'solution' contain this time?

_____

In which trial did you use more 'solute molecules' (objects)?

_____

In which trial did you have the least water? _____

Explain the relationship between the amount of solute molecules and the amount of solvent present in a solution. _____

_____

_____

Recall that molecules never stop moving, even after reaching equilibrium. Each of your cells has two solutions separated by a selectively permeable membrane – the extracellular fluid outside of the cell and the intracellular fluid inside the cell. These solutions usually have different compositions. If your cell membranes were fully permeable, all molecules could pass through and move down their concentration gradients until reaching equilibrium, and we would have simple diffusion. But the cell membrane is semipermeable. We know water can pass through the cell membrane rather easily, but many solute molecules can't because of their size or their chemical composition. You also know from your previous activity that there is a higher concentration of water (solvent) wherever there is a lower concentration of other stuff (solute molecules).

Look at **Figure 6.10**. In this illustration, there is a higher concentration of solute molecules inside the cell than there is outside. Assume the solute molecules shown cannot pass through the cell membrane. Now answer the following questions:

1. Which fluid contains the lowest concentration of solute molecules?

_____

2. Which fluid has the highest concentration of water? (Recall that when water is high, solute is low, and vice versa.)

_____

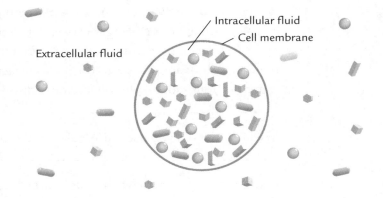

Extracellular fluid

Intracellular fluid

Cell membrane

FIGURE 6.10   **The cell membrane separates the extracellular fluid from the intracellular fluid.** In this figure, a situation is shown in which there are more solute molecules inside the cell than there are outside of it. Assume the shaded shapes are solute molecules and the remaining space is filled with water.

By osmosis, water diffuses – it moves down its concentration gradient – from where it is in higher concentration to where it is in lower concentration. You should see clearly from Figure 6.10 that, in this example, there is a greater concentration of water in the extracellular fluid than there is in the intracellular fluid, so water will enter the cell.

For osmosis, you just need to know where the highest concentration of water is, because it will always move from that area to the area with less concentration. The only reason to consider the other stuff (solute) is because it tells you where the water is – the area with the lower solute concentration has the higher water concentration, so the water will move away from that area.

Osmosis is diffusion of water through a selectively permeable membrane. ▪

What happens to cells when water moves across the cell membrane? **Figure 6.11** shows three views of red blood cells. Figure 6.11a shows a normal red blood cell, shaped like a biconcave disk. If water leaves a cell, the cell will shrink and the cell won't function efficiently. This is

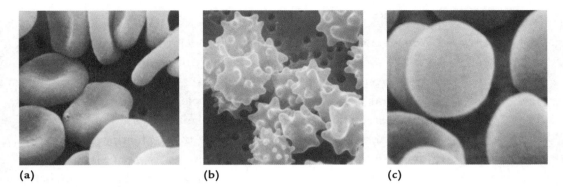

(a)   **(b)**   **(c)**

FIGURE 6.11   Red blood cells showing the effects of osmosis. **(a)** Normal red blood cells.
(b) Red blood cells that have lost water through osmosis. **(c)** Red blood cells swollen from
taking in too much water by osmosis, and at risk of rupturing.

shown in Figure 6.11b. If water moves into a cell, the cell will swell.
This is shown in Figure 6.11c. This swelling increases the pressure
inside the cell and impedes normal function.

LOOK OUT

Water moves into a cell through the selectively permeable
membrane and there is a point at which the cell resists further
entry – this is called **osmotic pressure**. **Osmolarity** is the total
concentration of all the solute particles in a solution. Learn these
terms, because they will help you understand what is happening to
your patient's physiological processes when their fluids are out of
balance, for example, when they may be dehydrated.

Under normal circumstances, the pressure of blood on the
capillary wall forces water out of the blood at the arterial end of
the capillary, but the solutes that are too large to pass out causes
fluid to be drawn back into the capillary as the blood pressure falls
at the venous end. Overall there is little loss of fluid into the
tissues.

PICTURE THIS

You are already familiar with water moving into and out of cells and changing their shapes and sizes. Unlike our cells, plant cells are surrounded by a rigid cell wall that helps maintain their shape. Have you ever had a houseplant or a plant in your garden that was a bit wilted and droopy? When the plant cells lose water, they shrink and cannot support the weight of the plant parts above them, so they droop. When they receive water, the cells swell, exerting pressure against the rigid cell wall, and this pressure helps support the rest of the plant – it stands back up.

APPLYING THE THEORY

Animal cells lack a cell wall, so they are more likely than plant cells to rupture if they take in too much water. In healthcare, IV (intravenous) fluids must have the appropriate amount of water – if there is more water in the IV fluid than in the patient's intracellular fluid, their cells will swell. If there is less water in the IV fluid, the patient's cells will shrink. Unless the patient is being treated for a fluid imbalance, IV fluids should have the same water concentration as intracellular fluids so the water will be at equilibrium. Such fluids are called *isotonic*. The wrong amount of water in an IV fluid can be fatal.

✔ QUICK CHECK

Refer back to Figure 6.10. Will water enter or leave the cell?

Answer: The water moves from where it is more concentrated to where it is less concentrated, so it moves *into* the cell.

Active Transport

So far we have discussed simple diffusion, facilitated diffusion and osmosis. These are all types of diffusion, so the molecules will move from an area where they are more concentrated to an area where they are less concentrated. Diffusion is a spontaneous process – it happens automatically. These types of diffusion do not require energy, so they are referred to as **passive transport**.

Sometimes, instead, cells need to move molecules against their concentration gradients. For example, for a nerve cell to transmit a signal, sodium ions must enter the cell. Special channels in the cell membrane open and sodium ions diffuse into the cell because there are more sodium ions outside than inside. This is simple diffusion. After that nerve signal is sent, though, the sodium ions must leave so the cell is ready for the next signal. But there are still more sodium ions outside than there are inside. The ones inside can't merely diffuse out – they would be moving up their concentration gradient.

PICTURE THIS

Assume you are a dedicated anatomy and physiology student and you've been studying much more than you've been keeping your house clean. You hear a van pull up outside and you see that it is the *Prize Patrol* from a major sweepstakes sponsor, with a camera crew. They are broadcasting live and they are quickly approaching your door. You grab an armfull of clutter and all your books and open a cupboard door to hide them on the overhead shelf, but the shelf is full because it has already been jammed with previous clutter.

1. What will probably happen if you just open the cupboard door and stand back? _____

2. Is it easy or difficult to cram more clutter onto the shelf?

   _____

3. Does the clutter spontaneously head onto the shelf on its own, or do you have to really work hard to get it to go and stay there?_____

In the situation with the cupboard, when you open the door there is more clutter inside on the shelf than there is outside, so the clutter inside will spontaneously fall out. This represents diffusion – molecules moving from where they are more concentrated to where they are less concentrated. You will probably have to use your arm or exert effort to prevent the clutter from falling out. At the same time, you have to force the additional clutter in your arms into the cupboard, and you have to work hard to do it. Work requires energy, so you must use energy to do it.

With cells, molecules will not spontaneously move against their concentration gradients. Work must be done, so energy in the form of adenosine triphosphate (ATP) must be used. For this reason, this type of movement – moving molecules up, or against, their concentration gradient – is called **active transport**. This type of movement also requires special one-way 'doorways' in the cell membrane, called **pumps**, that ensure the molecules can only move in one direction. Otherwise molecules on the other side would spontaneously diffuse out, in the opposite direction.

Active transport moves molecules against their concentration gradient, which requires a special molecular pump and energy. ■

Exocytosis

Cells may make secretory proteins and other large molecules that will be exported. These products are usually wrapped in a membranous sac, called a **vesicle**, at the Golgi apparatus. The method by which they are expelled is called **exocytosis** (*exo-* = outside, *cyto-* = cell) and it is rather simple. The vesicle makes its way to the edge of the cell and its membrane fuses with the cell membrane. As the vesicle pushes against the cell membrane, its own membrane ruptures and seems to peel back, becoming part of the cell membrane and releasing its contents into the extracellular fluid. **Figure 6.12** shows a cell secreting a product via exocytosis. Notice the contents spewing out of the cell.

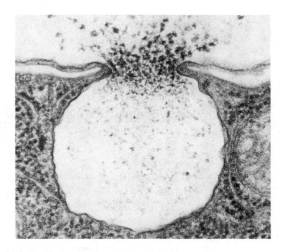

FIGURE 6.12 A cell performing exocytosis.

Endocytosis

Endocytosis (*endo-* = inside) is the reverse of exocytosis. It is a means by which cells can take in rather large objects or even liquid that contains dissolved materials, such as nutrients. There are three major types of endocytosis:

- phagocytosis,

- receptor-mediated endocytosis, and

- pinocytosis.

   Phagocytosis is the process by which solids such as bacteria or dead red blood cells are moved into your cells, and is sometimes referred to as 'cell eating'. This process demonstrates the active nature of the cell membrane. Extensions of the cell membrane, called **pseudopodia** ('false feet'), seem to reach out from the cell surface, rather like tiny arms, on each side of the object to be taken in. Then the extensions fuse and form a membranous sac, like a vesicle, around the object. This saclike structure is called a **phagosome**, and it moves inward, pinching off from the cell membrane on the inside of the cell. The object is now inside your cell. Soon, several lysosomes typically fuse with the phagosome. Some white blood cells in your immune system are specialised for phagocytosis and use this process to rid your body of foreign material, such as bacteria, that might make you ill.

✔ QUICK CHECK

Why do lysosomes fuse with the phagosome? _____

_____

Answer: Lysosomes contain enzymes that will break down the contents of the phagosome, rendering potential threats harmless and recycling materials that are then made available to the cell for reuse.

Receptor-mediated endocytosis is related to phagocytosis, but before the cell 'reaches out' to take the materials in, the objects to be moved must bind to special receptors in your cell membrane. This binding triggers the cell membrane to form a pocket that will enclose the materials and bring them into your cell. From this point on, the process is very much like phagocytosis.

APPLYING THE THEORY

Some enveloped viruses, such as the human immunodeficiency virus (HIV) and the influenza virus particle, have adapted their structure to fuse with the plasma membrane in order to enable the release of their genetic material into the cell cytoplasm. Here the virus particle can use the mechanisms of the cell to replicate itself. The fact that viruses can only survive as true intracellular parasites and not multiply on their own has led to some debate as to whether they are truly living organisms.

Pinocytosis is a way your cells can bring in liquids, and is sometimes referred to as 'cell drinking'. In this process, part of your cell membrane puckers inward, forming a pouch at the surface, and the pouch contains extracellular fluid. In that fluid are a variety of dissolved substances that are now surrounded by cell membrane, which pinches off to form a vesicle-like structure called an **endosome**. The contents are then released inside the cell and are available for its use.

Ashes to Ashes, Cells to Cells: **The Cell Cycle**

Remember from the Cell Theory that all cells come from pre-existing cells. Each cell goes through the **cell cycle** (**Figure 6.13**), and the duration of this cycle varies with the cell type. The cell cycle is divided into multiple phases. Simply, though, it can be viewed as having two main parts: **interphase** and **cell reproduction**.

Mature nerve cells do not divide, so they are always in interphase. Other types of cell have a variable time at interphase and hence length of cell cycle. For example, bone marrow cells have a cell cycle of 10–18 hours, whereas for red blood cells it is 120 days.

Most cells spend the majority of their life cycle in interphase, which is when they are doing their normal living activities. This makes sense – we spend most of our time living our life, and only a fraction, if any, in reproductive behaviour. During interphase, cells go about their normal business, growing, maturing and doing all the activities we have discussed and more. This is when the cell makes its contribution to the overall function of the whole organism.

During this phase, it also prepares for the reproduction to come. The DNA – your genetic material – replicates. That means it reproduces. Cells reproduce by dividing, but each daughter cell needs to have a complete set of all the DNA. So, during interphase, the complete set of DNA is copied in a process called DNA replication.

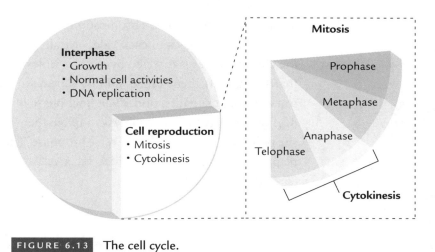

FIGURE 6.13   The cell cycle.

Once the DNA has replicated, cell reproduction may begin. Cell division, or reproduction, includes two processes:

■ **mitosis,** which is nuclear division, and

■ **cytokinesis,** which is cytoplasmic division.

The DNA is important to your cells, and division of the nucleus is a separate process, called **mitosis**. This is a very precise event designed to ensure that the DNA is copied perfectly and equally divided into the daughter cells. Mitosis includes four phases:

■ prophase

■ metaphase

■ anaphase, and

■ telophase.

During interphase, the DNA is stretched out in thin strands called chromatin. During **prophase**, the first phase of mitosis, the chromatin condenses into rod-like structures – the chromosomes (**Table 6.2**). The nuclear membrane also disappears so that the chromosomes can move more freely. The next phase is **metaphase**. *Meta-* means *middle*, and during metaphase the chromosomes align very precisely, in duplicated pairs, along the midline, or equator, of the cell. This ensures that they will separate precisely. The next phase is **anaphase**. *Ana-* means *away*, and the duplicated chromosomes separate during this phase – one complete set goes to each side, or pole, of the cell. Movement of the chromosomes is directed by the centrioles and delicate structures that form from them called spindle fibres, which pull the chromosomes in opposite directions. The final phase of mitosis is **telophase**, during which the chromosomes complete their journey to the opposite poles. This phase is like a reverse prophase. The nuclear membrane reappears and the chromosomes relax back into the stretched-out chromatin strands.

TABLE 6.2   **The stages of the cell cycle include interphase and cell reproduction.** Cell reproduction includes the four phases of mitosis during which the nuclear contents divide, and cytokinesis during which the remainder of the cell divides. Cytokinesis overlaps the latter phases of mitosis.

Picture	Stage	Events
	Interphase	Normal cell activities, growth and DNA replication. DNA is visible as thin strands called chromatin.
	Prophase	Chromatin condenses into rod-like structures called chromosomes that are clearly visible, and the nuclear membrane disappears.
	Metaphase	Chromosomes align very precisely along the midline.
	Anaphase	Chromosomes separate and are pulled to opposite poles of the cell. Cytokinesis begins once the chromosomes separate (visible here where the edge of the cell is beginning to pinch in).
	Telophase	Chromosomes are in opposite poles and cytokinesis continues. This stage ends when the cells completely separate, forming two daughter cells.

TIME TO TRY

In the spaces provided, list the phases of mitosis in the correct order, then sketch what the cell would look like during that phase.

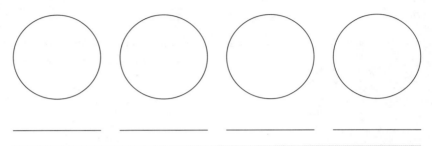

_____   _____   _____   _____

Mitosis just divides the nucleus or, more specifically, the chromosomes. Once the chromosomes have carefully separated, the rest of the cell can be divided. This process, by which the cytoplasm divides, is called **cytokinesis**. It begins during anaphase of mitosis, after the chromosomes have separated, and ends at the end of telophase. At the end of cytokinesis, the original cell is gone and in its wake are two new daughter cells, almost identical and each with a complete set of DNA. These daughter cells are in interphase, and the whole cell cycle begins anew.

Cell reproduction includes mitosis, which is division of the nuclear contents, and cytokinesis, which is division of the cytoplasm. ▩

APPLYING THE THEORY

Remember: both resultant daughter cells after cell division are almost identical. Each cell has a nucleus with a set of chromosomes from the original cell and a new set made during the cell cycle. **Mitosis** is for growth and tissue repair, so each cell must be similar. **Meiosis** is only for reproduction and so involves the making of the gametes (ova and spermatozoa) with resultant halving of the number of chromosomes in the nucleus (23 single), so that during fertilisation the normal 46 (23 pairs) is restored.

TIME TO TRY

Go to *Get Ready for A&P for Nursing and Healthcare* on your computer. Enter the web address **www.pearsoned.co.uk/getready**

Go to Chapter 6: Cell biology. Try the interactive tutorial on <u>Osmosis</u> and <u>Mitosis</u> and <u>Cytokinesis</u>.

Now do the <u>Post-test</u> quiz to find out what you have learnt in this chapter.

FINAL STRETCH!

Now that you have finished reading this chapter, it is time to stretch your brain a bit and check how much you have learned.

RUNNING WORDS

At the end of each chapter, be sure you have learned the language. Here are the terms introduced in this chapter with which you should be familiar. Write them in a notebook and define them in your own words, then go back through the chapter to check your meaning, correcting as needed. Also try to list examples when appropriate.

Cytology	Phospholipid	Protein synthesis
Cell	Phosphate head	Messenger RNA
Cell Theory	Hydrophilic	(mRNA)
Unicellular	Fatty acid tail	Ribosomal RNA
Multicellular	Hydrophobic	(rRNA)
Cell membrane	Phospholipid bilayer	Nucleolus
Cytoplasm	Selectively permeable	Transfer RNA
DNA	Fluid mosaic model	(tRNA)
Prokaryote	Cytosol	Endoplasmic
Eukaryote	Nucleus	reticulum (ER)
Nucleoid	Chromatin	Golgi apparatus
Plasmid	Chromosome	Vesicle
Organelle	Nuclear envelope	Mitochondrion
Homeostasis	Nuclear pore	ATP
Plasma membrane	Ribosome	Lysosome

Cytoskeleton	Osmosis	Pinocytosis
Microtubule	Solute	Endosome
Microfilament	Solvent	Cell cycle
Centrosome	Solution	Interphase
Centriole	Passive transport	Cell reproduction
Cilia	Active transport	DNA replication
Flagellum	Exocytosis	Mitosis
Brownian motion	Endocytosis	Prophase
Concentration	Phagocytosis	Metaphase
gradient	Pseudopodia	Anaphase
Equilibrium	Phagosome	Telophase
Simple diffusion	Receptor-mediated	Cytokinesis
Facilitated diffusion	endocytosis	

TIME TO TRY

Now look at the online Glossary in the *Get Ready for A&P for Nursing and Healthcare* website. Make a list of your new vocabulary and check out the meanings. You can also make flash cards to help you learn these new terms.

WHAT DID YOU LEARN?

PART A: ANSWER THE FOLLOWING QUESTIONS

1. List the five principles of the Cell Theory. _____

_____

_____

_____

_____

_____

2. What is the basic difference between prokaryotic cells and eukaryotic cells? _____

_____

_____

3. Describe the organisation of the cell membrane. _____

_____

_____

4. Design a concept map for the following terms: protein, DNA, nucleus, nucleolus, ribosome, rough ER, Golgi apparatus and exocytosis.

5. Differentiate between passive and active movement processes.

_____

_____

_____

6. In terms of movement processes, explain how making a cup of hot tea with a tea bag involves both osmosis and simple diffusion. _____

_____

_____

_____

_____

7. Assume you have limp carrot sticks in your refrigerator. To get them plump and crisp again, would you soak them in pure water or in a strong salt solution? _____

8. Differentiate between phagocytosis and pinoctyosis. _____

_____

_____

9. List in order the four phases of mitosis: _____

_____

_____

_____

10. Explain the complete cell cycle. _____

_____

_____

_____

_____

PART B: FOR EACH OF THE FOLLOWING ITEMS, MATCH THE TERM WITH ITS DESCRIPTION

1. mitochondrion _____
2. lysosome _____
3. ribosome _____
4. Golgi apparatus _____
5. nucleolus _____
6. pinocytosis _____
7. exocytosis _____
8. osmosis _____
9. active transport _____
10. simple diffusion _____

(a) Diffusion of water through a selectively permeable membrane.

(b) Expelling materials out of the cell.

(c) Site where ribosomes are made.

(d) Process that uses energy to move molecules against their concentration gradients.

(e) Spontaneous movement of molecules down their concentration gradient.

(f) 'Waste disposal' containing digestive enzymes.

(g) Site where proteins are made.

(h) Site where cellular products are packaged.

(i) Cell drinking.

(j) Site where ATP (energy) is made.

WEB RESOURCES

Here are some additional online resources for you.

■ *Get Ready for A&P for Nursing and Healthcare website*
Go to *Get Ready for A&P for Nursing and Healthcare* on your computer
Enter the web address **www.pearsoned.co.uk/getready**
Go to Chapter 6: Cell biology.

■ *BBC Education website – ASGURU biology*

http://www.bbc.co.uk/education/asguru

Look at Biology and then click on <u>Cell Biology</u> and <u>Pathways Into and Out of the Cell</u>.

■ *Nottingham School of Nursing*

http://www.nottingham.ac.uk/nursing/sonet/rlos/rlolist.php

Try clicking on <u>Cell Division</u> and also <u>Concentration Gradients</u> and <u>Osmosis and Diffusion</u>.

There are many other interactive tutorials here that may interest you.

■ *Cells Alive*

http://www.cellsalive.com

Try clicking on <u>Cell Biology</u>, <u>Mitosis</u>, <u>CellCams</u> and <u>CellCycle</u>.
This site offers many exercises and animations for exploring cell structure, mitosis and the cell cycle.

■ *Headstart in biology*

http://www.headstartinbiology.com

Gallery 4 starts with the **whole body**, and then looks at the named **regions** and the major **systems**. The next objective is to appreciate that all the structures of the body are created from and by **cells** – tiny living units that share many features but also become specialised for particular roles. Cells live together to form **tissues**, and tissues are assembled into distinct yet interconnected **organs**.
Look at this free website – you will need to register. There are five galleries. Gallery 4 is <u>Arrangement of the body</u>.
There are tutorials on each aspect and a quiz at the end of the gallery.

■ *Wisc Online*

http://www.wisc-online.com/default.asp

Sign up for free.
Try the Learning Activities: <u>A Typical Animal Cell</u>, <u>Construction of the Cell Membrane</u> and <u>The Cell: Passive Transport Diffusion</u>.

■ *Window On Life*

http://www.dti.gov.uk/publications/index.html

Search in Reports and Publications under W for **Window on life: my life**. A two CD-ROM set published as part of the 50th anniversary of the discovery of the structure of DNA. The set contains 260 articles by leading figures in genetics and bioscience, scientific authors, journalists and ethicists, with 400 images, curriculum links, educational materials, specially commissioned video featuring interviews with leading researchers, animations and activities which develop a background understanding of this complex field. Window on life: my life covers the development of genetics from the basic building blocks of life, through human evolution and ageing.
Order the two CD-ROM set for free.
Look at the animations <u>Protein Synthesis</u> and <u>Mitosis</u>.

Answer Key

CHAPTER 1
Answers will vary from student to student.

CHAPTER 2

Part A
1. −32
2. 20,000
3. 3/4
4. 1/4
5. 1/4
6. 0.2
7. 720 breaths per hour
8. 400 cm
9. 7500 mg
10. 45 kg

Part B
1. 14
2. 4
3. 0.3; 30%
4. 20%
5. volume of a liquid
6. gram; metre (technically kilometre, but in practical use it is the metre); litre
7. 'Normal' blood pressure is the average blood pressure.
8. 100°C
9. volume
10. 1000

CHAPTER 3

Part A
1. e
2. f
3. i
4. j
5. g
6. h
7. b
8. d
9. c
10. a

Part B
1. In the anatomical position, your palms are facing forward.
2. Sagittal.
3. d.
4. b.
5. Coronal and sagittal.
6. Polycythemia; hepatitis.
7. Abnormal narrowing of the arteries; cartilage cell.
8. Pharynges; mitochondria; coxae.
9. *Medial* is a comparative term meaning a designated object is closer to the midline; *median* means a structure is positioned exactly on the midline.
10. Oblique.

CHAPTER 4

Part A

1. Organism, organ system, organ, tissue, cell, organelle, macromolecule, molecule, atom.
2. The foot is a broadened surface with arches that distribute the weight evenly to all parts of the foot that contact the ground.
3. From the food we eat.
4. If you consume too much energy you gain weight; if you spend more energy than you consume, you lose weight; and if your energy expenditure equals your energy intake, your weight remains constant.
5. Homeostasis is the maintenance of a constant internal environment. It is important to maintain homeostasis because cells perform optimally in a balanced internal environment.
6. Epithelial, connective, nervous, muscle.

Part B

1. b, c
2. d, h
3. c, e
4. a
5. b, c, f
6. b
7. d
8. c, d, g

CHAPTER 5

Part A

	Potassium	Iodine	Oxygen	Neon
Chemical symbol	K	I	O	Ne
Atomic number	19	53	8	10
Atomic weight	39.10	126.90	16.00	20.18
Number of protons	19	53	8	10
Number of electrons	19	53	8	10
Number of neutrons	20	74	8	10
Number of electrons in outermost shell	1	7	6	8

Part B

1. Protons and neutrons.
2. Electrons.
3. Protons.
4. Carbon, hydrogen, oxygen, nitrogen.
5. The row tells you how many shells of electrons that element has.
6. The column tells you how many electrons are in the outer shell of that element.
7. The elements in the far right column are stable because they have a full outermost shell of electrons.
8. 2; 8.
9. The calcium atom lost two electrons, giving it a $^{+}2$ charge.

10. In ionic bonds, the electrons physically move from one atom to another so that the atoms involved lose or gain electrons, and the opposite charges of the resulting ions draw the atoms together. In covalent bonding, the atoms share electrons.

11. Structural.

12. Decomposition (catabolic).

13. Synthetic (anabolic).

14. Carbon and Hydrogen.

15.
$C_6H_{12}O_6$	O
CO_2	I
CH_4	O
CO	I
HCl	I
H_2O	I

CHAPTER 6

Part A

1. All organisms are composed of one or more cells; cells are the basic structural and functional units of life; all vital functions of an organism occur within cells; all cells come from pre-existing cells; and cells contain hereditary information that regulates cell functions and is passed from generation to generation.

2. Prokaryotes have no nucleus and lack internal membranes.

3. The cell membrane is a phospholipid bilayer with protein channels periodically passing through it and other molecules such as cholesterol and carbohydrates 'floating' in it.

4. Concept maps will vary from student to student, but these relationships should be included:
   - DNA contains the instructions for how to build proteins and is located in the nucleus.
   - The nucleolus is located inside the nucleus and makes ribosomes.
   - Rough endoplasmic reticulum gets its appearance from the presence of ribosomes.
   - Proteins are made at ribosomes, so rough ER makes proteins.
   - Proteins from rough ER move to the Golgi apparatus for processing and packaging, then are expelled from the cell by exocytosis.

5. In passive movement processes, substances move along their concentration gradients from high concentration to low concentration, e.g., diffusion and osmosis. In active movement processes (active transport), energy is used to move substances against their concentration gradients (from low concentration to high), e.g., exocytosis and endocytosis.

6. When a tea bag is placed in hot water, the water moves into the tea bag by osmosis and dissolves the tea, then the tea molecules diffuse out of the tea leaves into the water.

7. Water would move into the carrot, plumping the cells and making them rigid. (Salt would draw more water out, causing them to shrivel up even more.)

8. In phagocytosis, the cell surrounds a solid, with the cell membrane extending out to surround the object, then drawing it inward. In pinocytosis, the cell membrane forms an inward pouch that surrounds a droplet of liquid – bringing the liquid into the cell.

9. Prophase, metaphase, anaphase and telophase.

10. The complete cell cycle includes all events that occur in the life of a cell, specifically interphase and cell reproduction. Interphase is when the cell carries out its normal functions, and the DNA replicates in preparation for reproduction. Cell reproduction includes the four phases of mitosis that divide the nucleus, and cytokinesis that divides the cytoplasm.

Part B

1. j
2. f
3. g
4. h
5. c
6. i
7. b
8. a
9. d
10. e

Index

Photograph and Illustration Acknowledgements

All chapter opener art, cartoon spot art and Figures 3.5 and 3.6 was created by Kevin Opstedal. Where not credited elsewhere, all other artwork was created by Seventeenth Street Studios.

We are grateful to the following for permission to reproduce copyright material:

Learning styles assessment in Table 1.1: © Marcia L. Conner, www.agelesslearner.com; Table 1.2, *75118* © Getty Images, Inc; Figure 1.5 *O-074-0388* by Tom Stewart © Tom Stewart/Corbis; Figures 2.5 and 4.5c,e,f from figures 2.6 and 4.5c,e,f respectively of *Get Ready for A&P*, Pearson Benjamin Cummings, (Garrett, Lori K.); Figure 2.6a © Ohaus Corporation; Figure 2.7 *54330206A-2RM Meniscus: water in graduated cylinder* by Richard Megna © Richard Megna, FUNDAMENTAL PHOTOGRAPHS, NYC; Figure 3.3 photo of hip implant © National Institutes of Health; Figure 3.4 *The Reward of Cruelty*, Hogarth, William (1697–1764), © Courtauld Institute of Art Gallery, London; Figures 4.1, 5.1, 6.1 *neuron 10X* © Pearson Benjamin Cummings; *photomicrograph of motor PA0117738* by Ed Reschke © Ed Reschke/Still Pictures; *mitochondrion 900012* by Dr Don W Fawcett © Dr Don W. Fawcett/Visuals Unlimited; Figure 4.3a,b *228644* and *228640* by Dr Fredrick Skvara © Dr Fredrick Skvara/Visuals Unlimited; Figure 4.3c,d,e by R.T. Hutchings © R.T. Hutchings; Figure 4.4a *S6598A* by P. Motta © P. Motta/Photo Researchers Inc.; Figure 4.4b *SB7994* by Steve Gschmeissner © Steve Gschmeissner/Photo Researchers Inc.; Figure 4.5a by Nina Zanetti © Pearson Benjamin Cummings; Figure 4.5b *100X* © Pearson Benjamin Cummings; Figure 4.6 from *A Stereoscopic Atlas of Human Anatomy* (Bassett, David L., MD); Figure 4.7 and Table 5.1: adapted from figure 1.2 and table 2.1 from ESSENTIALS OF HUMAN ANATOMY & PHYSIOLOGY, 8th ed. by Elaine N. Marieb. Copyright © 2006 by Pearson Education, Inc. Reprinted by permission; Figure 5.8 *kittens photo* by Lori K. Garrett © Lori K. Garrett; Figure 6.2a *285806* by Dennis Kunkel © Dennis Kunkel/Visuals Unlimited; Figure 6.2b *SB4291 Purkinje nerve cell* by David McCarthy © David McCarthy/Photo Researchers Inc.; Figure 6.2c *P6573A* by David M. Phillips © David M. Phillips/Photo Researchers Inc.; Table 6.2 *The phases of mitosis in an animal: interphase, prophase, metaphase, anaphase, telophase and cytokinesis* by Ed Reschke © Ed Reschke; Figure 6.5 adapted from *Study Guide for Human Anatomy & Physiology, 6th Edition*, Pearson Benjamin Cummings, (Marieb, Elaine N. 2004); Figure 6.7 electron micrograph part, *9L8645 Mitochondria* by D.W. Fawcett © D.W. Fawcett/Photo Researchers Inc.; drawn part of Figure 6.7 adapted from *Microbiology*, Pearson Benjamin Cummings, (Bauman, Robert W. 2004); Figure 6.11a,b,c from *Journal of Cell Biology*, (Sheetz, M., Painter, R. and Singer, S. 1976); Figure 6.12 *Golgi Function-Exocytosis* by Dr Birgit H. Satir © Dr Birgit H. Satir, Ph.D.;

In some instances we have been unable to trace the owners of copyright material, and we would appreciate any information that would enable us to do so.